200 best aviation web sites
and 100 More Worth Bookmarking

More Aviation Titles from McGraw-Hill

The Illustrated Buyer's Guide to Used Airplanes, Fourth Edition
Bill Clarke

Airplane Ownership
Ronald J. Wanttaja

Aviation Computing Systems
Mal Gormley

The Aviation Fact Book
Daryl E. Murphy

*Beyond the Checkride:
What Your Flight Instructor Never Taught You*
Howard J. Fried

Flight Instructor's Pocket Companion
John F. Welch

*Airplane Maintenance & Repair:
A Manual for Owners, Builders, Technicians and Pilots*
Douglas S. Carmody

*The $100 Hamburger:
A Guide to Pilots' Favorite Fly-In Restaurants*
John F. Purner

200 BEST AVIATION WEB SITES

AND 100 MORE WORTH BOOKMARKING

John A. Merry

McGraw-Hill

New York   San Francisco   Washington, D.C.   Auckland   Bogotá
Caracas   Lisbon   Madrid   Mexico City   Milan
Montreal   New Delhi   San Juan   Singapore
Sydney   Tokyo   Toronto

Library of Congress Cataloging-in-Publication Data

Merry, John A.
    200 best aviation web sites: and 100 more worth bookmarking: unbiased reviews of the industry's finest Internet offerings / John A. Merry.
        p.   cm.
    ISBN 0-07-001646-1 (pbk.)
    1. Aeronautics—Computer network resources—Directories.
I. Title.
TL512.M37   1998   98-10909
025.06'62913—dc21   CIP

McGraw-Hill

A Division of The McGraw·Hill Companies

Copyright © 1998 by The McGraw-Hill Companies, Inc. All rights reserved. Printed in the United States of America. Except as permitted under the United States Copyright Act of 1976, no part of this publication may be reproduced or distributed in any form or by any means, or stored in a data base or retrieval system, without the prior written permission of the publisher.

1 2 3 4 5 6 7 8 9 0   FGR/FGR   9 0 3 2 1 0 9 8

ISBN 0-07-001646-1

Allen County Public Library
900 Webster Street
PO Box 2270
Fort Wayne, IN 46801-2270

The sponsoring editor for this book was Shelley Carr, the editing supervisor was Curt Berkowitz, and the production supervisor was Tina Cameron. Cover art and book design by Marla Meredith.

Printed and bound by Quebecor Fairfield.

All brand names and product names used in this book are trademarks, registered trademarks, or trade names of their respective holders.

Information contained in this work has been obtained by The McGraw-Hill Companies, Inc. ("McGraw-Hill") from sources believed to be reliable. However, neither McGraw-Hill nor its authors guarantee the accuracy or completeness of any information published herein and neither McGraw-Hill nor its authors shall be responsible for any errors, omissions, or damages arising out of use of this information. This work is published with the understanding that McGraw-Hill and its authors are supplying information, but are not attempting to render engineering or other professional services. If such services are required, the assistance of an appropriate professional should be sought.

 This book is printed on recycled, acid-free paper containing a minimum of 50% recycled, de-inked fiber.

Dedicated to my wife, who seems to have an unlimited supply of encouragement, creativity, and clever solutions.

Contents

Introduction 1

Review Rating Criteria 2

Online Updates 3

The Basics of Browsing 4

Aviation Directories
Smilin' Jack 8
Aviation Internet Resources 9
The Canadian Aviation Web 10
Captain Bob's Pro Pilot Page 11
DeltaWeb Airshow Guide 12
The American & Canadian Aviation Directory Online 13
AirNemo 14
Aviator's Reference Guide 15
VirtualAirlines.com 16
SpaceZone 17
Russian Aviation Page 18
WWW.NOTAM.COM 19
AERO.COM—Future of Aviation 20
The Air Affair Aviation Hotlist 21
TOTAVIA—Aviation Information Services 22
The Air Affair 23
Landings 24
WWW.FLIGHT.COM 25
The Aviation Home Page 26
AirNav 27
E-Flight Center 28
Cyberflight 29
R/C Web Directory 30

Contents

Charlie Alpha's Home Page 31
Aerolink 32
Aero-Web—Aviation Enthusiast's Corner 33
World of Flight (WOF) 34
Women in Aviation—Resource Center 35
Bookmarkable Listings 36
    Alex's Helicopter Home Page
    The Flight Deck
    Air Cargo Online
    R/C Airplanes NET
    AVLINK
    The Flying High Page
    Air Cargo Newsgroup Home Page
    Seaox Air-Medical Page
    Airlines on the Web
    The Pilot Pitstop
    AviationDirectory.com
    Isreali Airpark
    AviationNet
    Calin's Aviation Index
    Airshow.com
    Airship and Blimp Resources
    Aileron's Place
    Antique Aircraft Enthusiasts
    Army Aviation Directory
    Aviation Business Center

Aviation Organizations and Associations

Center for Advanced Aviation System
    Development (CAASD) 40
The Ninety-Nines—International Organization of
    Women Pilots 41
Women in Aviation International (WAI) 42

Contents

The National Transportation Safety Board (NTSB)........ 43
National Aeronautics and Space Administration (NASA) .. 44
Career Pilots Association (CPA)...................... 45
Air Force Link....................................... 46
The NATCA Voice...................................... 47
Cessna Pilots Association (CPA) 48
International Miniature Aircraft Association (IMAA) 49
Helicopter Association International (HAI)............ 50
AirLifeLine 51
Fly-In.org 52
American Institute of Aeronautics and Astronautics (AIAA) 53
World Flight 1997.................................... 54
The Official Blue Angels Home Page 55
USAF Thunderbirds 56
National Business Aircraft Association, Inc. (NBAA)...... 57
Aviation Safety Connection (ASC) 58
Aircraft Owners & Pilots Association (AOPA).......... 59
National Aeronautic Association (NAA) 60
MicroWINGS 61
Experimental Aircraft Association (EAA).............. 62
International Aerobatic Club (IAC) 63
Soaring Society of America (SSA)..................... 64
Federal Aviation Administration (FAA) 65
Bureau of Transportation Statistics (BTS)—
    Office of Airline Information 66
World Aeronautics Association (WAA) 67
International Wheelchair Aviators (IWA).............. 68
Air Transport Association (ATA)...................... 69
Angel Flight 70
The Mechanic Home Page 71

Bookmarkable Listings **72**
    The World League of Air Traffic Controllers
    469th Security Forces Home Page
    Vietnam Helicopter Flight Crew Network
    Naval Helicopter Association (NHA)
    Lindbergh Foundation
    United States Parachute Association (USPA)
    American Bonanza Society
    Mooney Aircraft Pilots Association
    United States Air Tour Association (USATA)
    Air Force Association (AFA)
    Institute of Navigation (ION)
    Civil Air Patrol
    Airport Net

Weather

The Weather Underground **76**
USA Today—Weather for Pilots **77**
National Weather Service **78**
Aviation Weather Center. **79**
WW2010—The Online Meteorology Guide. **80**
Weather—Cable News Network Interactive (CNN) **81**
American Weather Concepts (AWC) **82**
AccuWeather. **83**
Aviation Model Forecasts **84**
The Weather Channel **85**
WeatherNet: WeatherSites **86**
Weather by Intellicast **87**
Bookmarkable Listings **88**
    Singer's Lock
    The Weather Visualizer
    Real-Time Weather Data

Charles Boley's Weather Stuff Online
Atmosphere Calculator
National Climatic Data Center (NCDC)
Sunrise/Sunset/Twilight and Moonrise/Moonset

Pilot Resources

AirCharterNet 90
Fillup Flyer Fuel Finder 91
UK Airfields Online 92
Aviation Information Resource Database 93
FlightWatch 94
Equipped to Survive 95
TheTrip.com 96
Air Safety Home Page 97
The Air Safety Investigation Resource (ASI) 98
Aircraft Technical Publishers (ATP) 99
Aviation & Aerospace Medicine 100
Great Circle Distance Calculator 101
EarthCam 102
The Air Charter Guide (ACG) 103
Introduction to GPS Applications 104
Aero-Tourism 105
F.E. Potts' Guide to Bush Flying—Concepts and
   Techniques for the Pro 106
High Mountain Flying in Ski Country U.S.A. 107
The Hundred Dollar Hamburger—A Pilot's Guide to
   Fly-In Restaurants 108
AirPage—The Interactive Aircraft Handbook 109
Webflyer 110
FlightSafety International 111
GTE DUATS 112

CONTENTS

Bookmarkable Listings **113**
    Simcom Training Centers
    GoldenWare Travel Technologies
    Shareware Aviation Products
    Aerodynamics and Flight Simulator
    Exotic Aircraft Company
    Airwise Hubpage
    Interjet
    Best AeroNet
    Paul Tarr's GPS WWW Resource List
    The Homebuilt Homepage

Aviation Museums/Education/Flight Schools

National Warplane Museum **116**
Aviation Communication **117**
The Hangar **118**
Be a Pilot **119**
Applied Aerodynamics: A Digital Textbook **120**
Aviation Ground School **121**
SimuFlite Training International **122**
AirTimes **123**
The Aviation History On-Line Museum **124**
See How It Flies **125**
Air Forces of the Americas Almanac **126**
Helicopter History Site **127**
TheHistoryNet Archives—Aviation and Technology .. **128**
US Air Force Museum **129**
National Air & Space Museum (NASM) **130**
Aviation Web **131**
San Diego Aerospace Museum **132**
Neil Krey's Flight Deck **133**
The Aviation History of Wichita, Kansas—The Air Capital . **134**

University of Nebraska at Omaha (UNO) Aviation Institute 135
A Virtual Museum Describing the Invention of the Airplane 136
FirstFlight 137
Embry-Riddle Aeronautical University 138
Amelia Earhart 139
TIGHAR 140
Dryden Research Aircraft 141
Learning to Soar 142
Bookmarkable Listings 143
     Port Columbus Historical Society
     106th Rescue Wing
     390th Bombardment Group
     The College Aviation Resource Page
     Virtual Tour of the Boeing 727 Cockpit
     GG-Pilot
     Aero Data Files
     UND Aerospace
     History of the Brazilian Air Force
     College of Aeronautics
     Aeroflight
     Luc's Photo Hangar
     8th Air Force in World War II
     Rhinebeck Aerodrome Museum
     Air Cruise America
     The International Women's Air & Space Museum
     CFI Central
     The Spruce Goose
     American Airpower Heritage Museum
     The Air Base

Aviation Online Magazines & News
CyberAir Airpark 148

CONTENTS

The Homebuilder's Den **149**
AeroWorldNet **150**
Aviation Weekly **151**
General Aviation News & Flyer **152**
Seaplane Pilots Association **153**
Rotorcraft.com **154**
Business & Commercial Aviation (B/CA) **155**
FlightWeb **156**
A/C Flyer Online **157**
Pilot's Web **158**
Aviation Safety Web Pages **159**
Jane's Information Group **160**
Ultralight Flyer Online **161**
FlightLine **162**
Airfax **163**
US Aviator **164**
AVWeb **165**
Air Chronicles **166**
Aviation Digest **167**
Aviation From Pilot—the UK GA Magazine **168**
Aviation & Aerospace by McGraw-Hill **169**
Inflight USA Magazine Online **170**
Internet Business Air News **171**
The Professional Aviator **172**
Journal of Air Transportation World Wide (JATWW) .. **173**
The Aviator's Hangar **174**
The Avion Online Newspaper **175**
Air & Space Home Page **176**
Bookmarkable Listings **177**
    The Controller
    FsFan BBS
    UK Airshows Review

Aeronautx
GPS World Online
Aerocrafter
Skydive!
Aviation Disasters
Flight Forum
United Space Alliance
National Championship Air Races

Aviation Parts/Supplies/Aircraft

Aviation Industry Resource (AIR) 180
Aerosearch 181
VisionAire Corporation 182
Aviation Café 183
AvShop.Net 184
Spinners Pilot Shop 185
WWW.Plane-World.com 186
Aeroprice 187
McDonnell Douglas (MD) 188
Bombardier Aerospace Group 189
Northrop Grumman Corporation 190
Lockheed Martin Corporation 191
Microsoft Flight Simulator 192
The Aviation Online Network 193
Global Aviation Navigator 194
PC Aviator 195
Aircraft Shopper Online (ASO) 196
Air Source One 197
Jeppesen 198
Wings Online 199
Optima Publications 200
AirShow—Aviation Trading Network 201

Contents

The Official Site for Learjet, Inc. **202**
Raytheon Aircraft **203**
Boeing **204**
The New Piper Aircraft, Inc. **205**
Bookmarkable Listings **206**
    Rockwell
    Europa
    Avsupport Online
    U.S. Wings Aviation Mall
    WSDN Parts Locator
    Internet Parts Locator System
    007 Aircraft Classifieds Online
    Web Wings Ltd.
    Airbus Industrie
    Airhead Pilot Shop
    Aircraft Suppliers Company

Aviation Entertainment

The Flight **210**
Dave English's Web Site—Great Aviation Quotes **211**
The Mile High Club **212**
Plane Spotting **213**
The World of Aviation Poetry **214**
Dave, Carey and Ed's Lancair Super ES Kitplane
    Progress Page **215**
Virtual Horizons **216**
From Buffalo to Alaska **217**
Captain J's Aviation Page **218**
Flying Contraptions Home Page **219**
Solo Stories **220**
Discovery Online—Wings Conversations **221**
Fudpucker Airlines **222**
Paper Airplanes by The PC Help Group **223**

"Dad" Rarey's Sketchbook—Journals of the 379th Fighter . **224**
Aviation Jokes **225**
Bookmarkable Listings **226**
    Air Pix Aviation Photos
    Tom Claytor—Bush Pilot
    Greg's Common Commercial Aircraft Spotter's Guide
    The Spotter's Homepage

Aviation Employment

Aviation Employee Placement Service (AEPS) **228**
Air, Inc.—The Airline Pilot Career Specialists **229**
Aviation Jobs Online **230**
www.FindAPilot.com **231**
Airline Employment Assistance Corps (AEAC) **232**
Your Career in Aviation—The Sky's the Limit **233**
Aviation/Aerospace Jobs Page (NationJob Network) **234**
Bookmarkable Listings **235**
    AeroTrek
    The Corporate Aviation Resume Exchange
    Universal Pilot Application Service
    Airline Pilot Job Update

Index **236**

Introduction

Combing the Web for worthwhile aviation sites may leave you cringing at the clock. The limitless selection can easily be overwhelming, not to mention time consuming. Even fishing with your favorite search engine oftentimes reels in every site except the one you want. Enter *200 Best Aviation Web Sites . . . and 100 More Worth Bookmarking.*

The following pages represent over 3,200 hours of browsing and reviewing. Here you'll find award-winning sites ranging from aviation news to pilot resources, hand-picked by pilots for pilots. The book's purpose, of course, is to help you avoid a tedious cyber-hunt enroute to aviation's better Web sites. And, rest assured that the choosing was performed in a completely unbiased way. Nothing listed here appeared as a result of paid advertising or other favorable treatment. If we thought the site to be worthy of your time, we've included it.

Review Rating Criteria

As with most reviews, the author's subjectivity ultimately becomes the predominant rating criteria. However, knowing that this won't fly with most aviation enthusiasts, I've established a few more tangible guidelines against which the *200 Best Aviation Web Sites* were judged:

Content: Did I uncover practical data and substance or a cesspool of typos and blurry plane pics?

Layout/Design: Was I bored to tearful yawns or mystically enthralled with site esthetics?

Functionality: Is site navigation a frustrating maze of futility or a wondrous example of efficiency?

Overall Audience: Does the site offer benefits to twelve people or a million and twelve?

Scale:

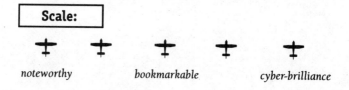

noteworthy   bookmarkable   cyber-brilliance

Online Updates

While each site's address and content has been checked (and rechecked), please bear in mind that addresses and page info may change or evaporate completely with time. It's simply the nature of the beast.

To keep up with aviation's dynamic sites, however, you are invited to check into *200 Best Aviation Web Sites'* Online Updates page for the latest in additions, deletions, and address (URL) changes. Stop in by pointing your browser to:

http://www.200bestaviation.com/updates

Each of the 200 sites found in the book, as well as the 100 bookmarkable listings, will be continuously monitored for changes and reported to you via the Online Updates page. Whether you're having trouble accessing a site, or just want a list of site modifications, give the page an occasional visit. It may save you some frustration.

The Basics of Browsing

Although the following may be old news for seasoned surfers, new Internet users may appreciate a quick intro on Web browsing basics. If you fall into the rookie category, welcome to the Web.

First, some terminology:

Browser—A software program used to view and navigate Web pages and other information. The most popular browsers include Netscape Navigator and Microsoft Explorer. We recommend using the latest versions of either of these two browsers, as most Web sites are formatted specifically for them.

URL—A Uniform Resource Locator (URL), also referred to as a Web site address, points to a specific bit of information on the Internet. For example, "http://www.200bestaviation.com" is a URL.

Bookmarks/Favorites—Most browsers offer a convenient way of storing and organizing your favorite Web sites in the form of a "bookmark" or "favorites" index. Adding a site to a bookmark/favorites list saves you the effort of retyping the site's URL during future visits.

Plug-In—As you access some sites in this book, you may notice the need for "plug-ins," or helper applications, to run video or audio features on the site. When using the latest versions of Netscape Navigator or Microsoft Explorer, you'll discover most plug-ins are already installed. However, if you find you're lacking a specific plug-in, chances are you'll be able to download it free. Most Web sites will provide a link back to the plug-in download site.

Second, armed with this new Web knowledge you're ready to begin browsing. Simply type the site address into your browser and press "enter." Be sure to pay particular attention to any special characters or upper case letters in the URL. To access each site, you'll need to type the address *exactly* as it appears in this book.

Third, be patient. Unforeseen forces sometimes determine your success at bringing up a site. It may be temporarily down. The actual communication lines may be jammed with too many users during peak times. Or, the site simply has withered away. Our advice? Try retyping the address another day, and move on to a site that does work.

So go on now, type, click, and bookmark.

Aviation Directories

AVIATION DIRECTORIES

Smilin' Jack

http://www.smilinjack.com

RATING
✈ ✈ ✈

BRIEFING:
A quality aviation/airline directory that keeps you "smilin'."

Thoughtfully void of giant pics and silly graphics, this sleek aviation directory creates minimal drag and maximum lift. As noted on the page itself, it was designed for easy loading and navigation. A simple menu guides you into efficiently organized content for: airlines (a huge list of airline links), fun flying (hyperlinks to popular sites), servers (more links—mostly directory sites), airports (comprehensive list of major international stuff), shopping, home pages (miscellaneous smattering of favorites), and weather (over 25 sites make you weather-wise).

By the way, in case your inquisitive mind really wants to delve into the background of Smilin' Jack, here's the scoop: he was a newspaper cartoon character created by Zack Mosley that ran from 1933 through 1973.

Fee or Free: Free.

DIRECTORIES

Aviation Internet Resources

http://airlines-online.com

RATING

BRIEFING:
Aviation subject searching doesn't get much better than this—really!

Tired of wading through unproductive online searches that cut into flying time? Well, roll your clicker into this breath of fresh air for to-the-point info. The folks at Aviation Internet Resources invite you to simply enter some pointed keywords for aviation-only database searches.

The efficiently designed index page employs conveniently clickable topics for: Airlines—the airline Web site database; Airports—the airport Web site database; Aviation—the aviation Web site database; Add a Site to Database; Daily Aviation News Briefs—searchable topic list; Online Discount Reservations; and Advertising Info.

If you're a traveler (hey, you are an aviation enthusiast . . .), I urge you to sample the Internet Travel Network online reservations system—it's well done and free! Take a trip through the online demo to discover: a secure online reservations system, low fare search mechanism, traveler's profile feature, and more.

Fee or Free: Free.

AVIATION DIRECTORIES

The Canadian Aviation Web

http://www.cavok.com

RATING

BRIEFING:
Canadian aviation—everything: databases, searches, links, news, and more from obvious Web pros.

There is a reason why this Canadian aviation site's an award-winner. Just scan through the enormous list of info-rich topics and you realize why it's Canada's leading online aviation resource.

Now maintained by TOTAVIA Aviation Information Services, The Canadian Aviation Web serves up aviation info through an expertly maintained, visually delightful collection of pages. Usually found in masterful sites such as this, a convenient "go to" pull-down menu combines with a margin list of clickable contents to unleash noteworthy navigation. Specifically, you'll find a huge list of aviation links searchable by province or scope/topic, newsgroups & forums, classifieds, news, events, fuel listings (current prices by province), aviation safety, learn to fly, fear of flying FAQ, and more. You'll also stumble across a healthy list of sponsors. It's no wonder why these companies clamor to be involved with The Canadian Aviation Web.

Fee or Free: Free.

Captain Bob's Pro Pilot Page

http://www.propilot.com

RATING
+ + +

BRIEFING:
A simple, easygoing collection of aviation stuff for your viewing and listening pleasure.

Captain Bob's Pro Pilot Page isn't about giant, slow-loading plane pics. Nor is it particularly fancy with modem-stalling graphics. It is, however, a solid, simple resource for pilots and aviation enthusiasts. Period.

Captain Bob (a California captain for Skywest Airlines), with his cyber-enthusiasm, has amassed a nice collection of aviation-related links and resources. The best parts? The page doesn't take forever to load, it's free of memberships, and all the hyperlinks seem to work without fail.

When you can catch your breath from all those giant, time-wasting aviation sites, check in here for: hordes of aviation links, airline/airport-specific sites, weather resources, a pilot bulletin board that's full of entries, an interesting commentary on regional airline careers, background on Captain Bob, lots of live video cams of popular destinations, and more.

My personal favorite: the live links to DFW Air Traffic Control and Chicago Air Traffic Control. To tune in, though, you will need a RealAudio player or equivalent. Most newer browser versions have one built right in.

Fee or Free: Free.

AVIATION DIRECTORIES

DeltaWeb Airshow Guide
http://www.deltaweb.co.uk/asgcal

RATING
⊹ ⊹

BRIEFING:
Conscientiously updated worldwide guide to airshows.

As noted in the Guide's text descriptions, airshows by their very nature are apt to change—sometimes radically—at the last minute. DeltaWeb's Airshow Guide does a great job combining current updates with well-oiled page organization.

The calendar is arranged by region (as of review time regions include the UK, North America and Europe, and the rest of the world) and further subdivided into months. Simply find your region and pick a month. Airshows will be listed in date order with location, short descriptions, and contact info. For example, a June airshow listing might read: "Saturday 7, Scott AFB, Illinois. HQ Air Mobility Command Base Open House and Airshow USAF 50th and Salute to the Berlin Airlift, Thunderbirds and Golden Nights. Tel: USA 618-256-1663."

Navigation throughout is painless. From each page additional months for the same region are readily available via your clicking fingertips. Conversely, other regions for a chosen month are standing by simply with a click of the regional icon.

Fee or Free: Free.

AVIATION DIRECTORIES

The American & Canadian Aviation Directory Online

http://www.infomart.net/av

RATING

BRIEFING:

Complete American and Canadian aviation site directory that lists, searches, and categorizes without charging a penny.

Bursting at its cyber-seams, the American & Canadian Aviation Directory is a one-stop resource with over 32,000 aviation companies listed. Conveniently available via a custom site searching tool, this aviation-only smorgasbord launches you into online aviation with only a couple of clicks.

Once you've maneuvered around the site's miscellaneous banners, your search form should begin to come into view. Simply enter any or all pertinent query data—from category to telephone number—and search. The directory does the rest. After it chunks through a fairly complete database on the other end of your modem, up pops your results. Category curious? Here are the highlights (with a lot of other stuff in between): aircraft charter/nonscheduled air transportation; aircraft dealers and brokers—retail; airports, flying fields, and airport terminal services; flight training; airports and aircraft maintenance; aircraft finance; aircraft storage; and more.

Itching to get your own aviation site noticed? Add your Web site address and company details here—it's free. Hey, when did you last read the words "free" and "aviation" in the same sentence? That's what I thought. Take advantage!

Fee or Free: Free.

AVIATION DIRECTORIES

AirNemo

http://www.geocities.com/CapeCanaveral/4285/index.html

RATING

BRIEFING:
An airline-info diamond in the rough that's hard to find using traditional search engines. You'll be glad I did the digging.

It's never late. Departures are always on time—24 hours a day. You won't even need to sprint to the gate. Just grab a mouse and settle into a first-class seat. Destination? AirNemo—The Best Links to Airline Sites. Dedicated to the air transportation industry, AirNemo boasts a knowledgeable site author behind the scenes who works for an international airline in Brussels, Belgium. As such, you might have guessed that French descriptions are conveniently standing by if English won't do.

Stylishly unobtrusive graphics and icons dominate the visual horizon, while excellent organization makes for frustration-free navigation. Look for convenient site codes like "new" flags, "official/unofficial" designations, and more. And, tying the pages together with invisible packing tape is a nice omnipresent, bottom-page frame menu.

At first glance, you may think that airline links and stats make up the entire site contents. While it's true, you'll find commercial aircraft characteristics about the B747-400, A340-200, and MD-95, and others, I urge you to dig deeper. The corporate aircraft characteristics, acronyms, conversion tables, links, sites, and codes are fascinating. Even if you disregard all of the above recommendations, do tap into the weekly Air Bulletin. It's simply a well-done online gem for aviation, air travel, and related issues.

Fee or Free: Free.

AVIATION DIRECTORIES

Aviator's Reference Guide

http://www.page4life.nl/aviator/options1.htm

RATING

BRIEFING:

Unprecedented aviation searching tool devoted to an earth-shattering list of reference links.

It's such a simple aviation searching breakthrough. Yet, the Aviator's Reference Guide seems to be the pioneer. What is this simple, yet overlooked aviation Web wonder? It's an omnipresent menu bar at the top of your screen used to quickly move between aviation topics of interest. Once you've chosen and clicked into an aviation reference link, such as N-registered Aircraft or The Airman's Information Manual, the reference topic menu stays with you wherever you go.

Always at hand, the unobtrusive bar menu takes up almost no screen space, but speeds searching time 100-fold. Done working with the Airport Data by Name/Code Query? Move on to, say, Aviation Headline News by clicking the pull-down, quick-link list of topics. If needed, "back" or "forward" options are also at the ready.

Okay, the searching tool is cool, but does the Aviator's Reference Guide contain the juicy aviation data you're always craving? You be the judge. Tap into: Airport Web Sites, The FARs, Meteorology Guide, Weather Abbreviations, Pilot Air News, Professional Pilots Rumor Network, Airline Directory by Country, Pilot Supplies, and way more topics than I have room to list.

Fee or Free: Free.

200 Best Aviation Web Sites **15**

AVIATION DIRECTORIES

VirtualAirlines.com

http://www.virtualairlines.com

RATING
✝ ✝

BRIEFING:
A handy hub dedicated to the fanciful world of virtual airlines.

Fee or Free: Free.

New and needed as a central virtual airline hub, VirtualAirlines.com boldly presents its "terminal" of information. With the virtual airline (VA) industry evolving into a complex maze of airlines and offerings, VA.com's gateway makes the search for VA-related companies and news an easy one.

The popular business of running a virtual airline involves flying routes, following procedures, promotion, and more. Without the expertly organized terminal guide of VA.com, the pilots, operators, and passengers would find themselves hopelessly lost.

Brushing under the tarmac a few description typos, the site expertly organizes and informs its stopover surfers with brevity and style. Esthetics, in the form of well-used graphics and layout add to the experience. But, at the core, the "terminal" menu ultimately gets you to your destination. Clickable gates include: General Forum & Site News Center (under construction at review time), Virtual Organizations, Information Center, Events, Special Offers, Membership Center, Worthwhile Links, and of course, quick jumps to Virtual Airlines. VA carriers currently operating from "Concourse C" include: AMEAST Airways, AviaStar, Montana Express, Northwind International, EuroStar, and many more.

Visit VirtualAirlines.com. There's no waiting. The terminal parking's free. And, the flights are always on time.

AVIATION DIRECTORIES

SpaceZone

http://www.spacezone.com

```
RATING
+ + + + +
```

BRIEFING:

A masterful mix of multimedia gizmos and futuristic design for space-age enthusiasts.

Almost as infinite as its subject, the SpaceZone site touches down upon an unending collection of goodies. News, events, history, and education certainly create a solid core. But, the true site mission revolves around a multimedia mix of unearthly content and out-of-this-world presentation. Contained within an electrifying display of design, fun gadgetry includes live NASA audio/video and coverage of space-related news, events, missions, and more. The true gravity of this site's excellence takes hold with topics like MIR Space Station news, NASA Television updates, and Space Shuttle happenings. Get the latest headlines or view the most recent series of video. For an uninterrupted experience, though, be prepared with your multimedia browser plug-ins of: VDOLive Player, RealAudio, and QuickTime (or their equivalents). Not plug-in savvy? Don't scrub the launch. SpaceZone conveniently offers the downloading links—remember they're free.

Once you're up on the most recent reports and rendezvous, go back a few light years into a complete look at space history. With stunning visual images and rare NASA footage, SpaceZone captures the excitement of exploration. Get space privy with astronaut biographies, space history timeline, mission footage & photographs, personal accounts, and planetary probes.

Fee or Free: Free.

This heavenly compilation is enough to make you weightless with space-driven ecstasy. If the page design doesn't launch you, the content will.

AVIATION DIRECTORIES

Russian Aviation Page

http://aeroweb.lucia.it/~agretch/RAP.html

RATING
✈✈✈✈

BRIEFING:
Russian aviation info riches o' plenty. Enjoy your flight.

Getting globally gregarious, I've accidentally stumbled across one of the world's most exhaustive informational sites on the Web today. The Russian Aviation Page is utterly bottomless in depth and pleasantly surprising in breadth.

Faster than a MiG-21 flyby, a flurry of well-designed aviation info bursts onto your screen from the moment you type the address. Even if Russian aviation stats and features aren't high on your bookmark list, I nevertheless encourage a visit. There's news, Russian aviation FAQs, museum review, digital movies, image archive, monthly Russian aviation trivia, and countless links.

Although page contents are carefully arranged and presented with award-winning style, there's almost too much to digest in one modem visit. Plan accordingly. Multiple menus, helpful search engines, and "new/update" tags offer the best assistance as you maneuver through this Russian wonder.

Skimming the Soviet surface, some must-see features include: chronology (development of the aviation industry in Russia and Soviet Union: 1916–1946), Soviet version of Top Gun, Soviet Superfortress—the Tu-4 Story, and many downloadable movies.

Intrigued by Soviet sites? Researching Russian aviation? Stow away on a Tu-142 and join other enthusiasts here.

Fee or Free: Free.

AVIATION DIRECTORIES

WWW.NOTAM.COM

http://www.notam.com

RATING

BRIEFING:
Interested in hang gliding, commercial, and general aviation? Here's your one-stop shop.

OK, it's a little bit all over the map. Smooth out your sectional and plan a course through this diversified directory. Tap into civic and FAA authorities, along with airlines, job opportunities, poems, airshow schedules, stories, images, and more. Specifically, you'll uncover: airlines on the Net, an aviation cybermall, flight simulation, flying clubs, weather links, and more. Then, stray from the straight and narrow with basic radiotelephony, conversion tables, the aviation disaster page, and visual & pyrotechnics signals.

For those looking for more one-on-one connections, your clickable choices include: a huge variety of newsgroups, pen pals, resume connection, corporate aviation, resume exchange, discussion forums, chat room, and post office.

Kick your feet up and explore here—there's something for everyone.

Fee or Free: Free.

AVIATION DIRECTORIES

AERO.COM—Future of Aviation

http://aero.com

RATING

BRIEFING:
Have patience while loading—a worldwide reference resource of magazines, newspapers, and newsletters are worth the standby time.

Yes, you'll have to wait a few seconds for the main menu to load—but it's worth it. This nice little directory of links features a couple of extras you wouldn't normally find: ballooning, parachutes, and a "helicoptorial," for instance. Step into the online publishing library for an eye-opening list of magazines, newspapers, and newsletters. The AERO.COM publishing page features a fairly comprehensive software catalog; nice collection of books; many mail order catalogs; and the classified ad section offers up a wide range of aviation toys.

You can also tap into the FAA Online for *NorCAL, SoCAL, Hi Desert Airman, Pacific Island Flyer,* and The Aviation Yellow Pages.

Fee or Free: Free. But, if you're subscribing to any pubs here, or plan to make any catalog purchases, you'll need to register (also free).

AVIATION DIRECTORIES

The Air Affair Aviation Hotlist

http://www.airaffair.com/hotlist.html

RATING
++

BRIEFING:
A fast, not pretty, guide to every aviation link you could possibly dream up.

Talk about your no-frills cheap seats. The Air Affair brings economy class straight to your monitor. Yes, it's part of the Air Affair site, but there's so much it deserves its own space.

If you're the kind who's annoyed by giant, blurry plane pics—here's the directory for you. Over fifteen broad categories and nine more in the Aviation Marketplace deftly move you into a comprehensive list of links. Here's a sampling (take a breath first): flying clubs, flight schools, FBOs, airports, publications and periodicals, homebuilt and amateur aircraft, charter and tour operators, government and regulation, aviation services, aircraft dealers, software, accessories and retailers, model and hobby aircraft, and more.

Bookmark this outstanding list site. It's excellent for finding anything from GTE DUATS to AeroCrafter—homebuilt aircraft sourcebook.

Fee or Free: Free.

AVIATION DIRECTORIES

TOTAVIA–Aviation Information Services

http://www.totavia.com

RATING

BRIEFING:
Nifty little directory site is a Canadian aviation catchall.

Canadian? Cool. Brought to you by the folks up north, TOTAVIA gives you a compact site of variety. You'll run into stuff like: Canadian aviation news (bump up your browser font size—it's a bit small), photos, and weather. Although you might experience uncomfortable loading times, site highlights include: Canadian airshows online, as well as links to US & UK events and airshows; aviation forums and online chatting; and a huge photo archive comprised of individual collections.

Compensation for the sometimes slow server comes in the form of handy navigation. A great search engine, links with descriptions, and pull-down menus help unleash the overwhelming volumes of Canadian aviation info.

Fee or Free: Free.

AVIATION DIRECTORIES

The Air Affair

RATING
✈ ✈ ✈ ✈

BRIEFING:
An online directory that gives you stunning efficiency by people who know what they're doing.

It's always refreshing to find an aviation site that truly has your interests in mind. The Air Affair does have the news, photos, links, and event listings that most "e-zines" have. But once you begin maneuvering through the selections, you notice the distinct difference of complete organization and speed. Put another way, VORs work fine most of the time, but it's nice to just punch the GPS a few times and get to your destination with ultimate ease. The Air Affair features: designed to be browser-independent (nice); lack of frivolous graphics (not necessarily a beautiful piece of Web art—but it's fast); thoughtful searching engines for fuel prices; local flight training locator; and well-thought-out page navigational aids.

Bookmark this site for the ultimate in: events, aviation fuel prices, news and views, aviation library, flying destinations, flight training locator, aviation links, hall of fame, and photo gallery.

Fee or Free: Free.

AVIATION DIRECTORIES

Landings

http://www.landings.com

RATING

BRIEFING:
Award-winning directory/database for every aviation subject imaginable.

Dubbed "aviation's busiest cyber-hub," Landings guides you into a huge, expertly maintained collection of info goodies. Sophisticated search engines move you through FAA and Canadian regulations, AIM, the pilot/controller glossary, service difficulty reports, airworthiness alerts, NTSB briefs, N-numbers, the FAA Airman Database, airman knowledge test info, and more. There are good links all over the place. Directory subjects include: aerobatics/flying, aircraft sales, aircraft-service/parts, manufacturers, airlines, airports, aviation BBSs, aviation images, aviation news groups, avionics, companies, flight schools/FBOs, flight planning, general aviation, GPS/technologies, hang gliding/paragliding, helicopters/gyrocopters, homebuilding, military, museums/history, publications, soaring, travel tours, and worldwide weather links that would leave goose bumps on a meteorologist.

The best part of this enormous informational grab bag is at-a-glance navigation. You'll find the concise "front page" offers news and quick-click menu categories.

Check in here often. They maintain and update regularly.

Fee or Free: Free, but do sign the log book for special offers and invites.

AVIATION DIRECTORIES

WWW.FLIGHT.COM

http://www.flight.com

RATING

BRIEFING:
Plain Jane site that wastes no time in finding you solid aviation resources.

It's simple. But sometimes tooling around in a Cessna 150 is kinda fun. Four topics grace the page: product directory, FARs, Hawk FTP site, and links. Use the product directory to find anything by name or company, or examine the FARs—they're all here. Download handy stuff like specific aircraft checklists, jobs, and news with Hawk FTP.

The somewhat disorganized links are fun if you have the time.

Fee or Free: Free.

AVIATION DIRECTORIES

The Aviation Home Page

http://www.avhome.com

RATING
✈ ✈ ✈ ✈ ✈

BRIEFING:
One click and you'll see why it's an award-winning aviation directory. Intelligently organized categories are completely searchable. It easily speeds you to your destination.

Just when you thought there were enough aviation directories... surprise! Here's another one. But, this award-winning site is simple and searchable. Completely redesigned and current, the Aviation Home Page gives you a nice summary BEFORE you click on subjects. Great idea—I wish more directories made it this easy. Sift through: airlines and airports; clubs, organizations, and companies; academies, universities, flight schools, and FBOs; federal and military resources; newspapers, magazines, schedules, and museums; flight simulation; art, photography, and poetry; weather, meteorology, and satellite images; classifieds; site news; and more!

Before departure be sure to look into their Select Sites that focus on the better aviation sites. Just choose a category and find your favorite.

Fee or Free: Free.

AVIATION DIRECTORIES

AirNav

http://www.airnav.com

RATING

BRIEFING:
A virtual smorgasbord of free airport, navigational fix, and fuel data.

OK. Enough flying frivolity. Let's get down to some serious navigational aids. AirNav pulls critical flight planning data out of cumbersome FAA publications and displays it on a virtual silver platter. Free and at your fingertips, you'll breeze through current facts, figures, and frequencies. Click on Airport Info for an amazing array of searchable airport data (way more than what you would find in the Airport and Facilities Directory). Navaid Info gets into the nitty-gritty about radio navigation (VORs, NDBs, TACAN, marker beacons, etc.). Fix Info provides enroute fixes, airway intersections, and waypoints ("From AAAMY to ZZAPP"). Then, figure in your pit stops with the Fuel Stop Planner.

It's a well-organized site, to begin your well-organized flight planning.

NOTE: *It's important to remember that info contained here is NOT valid for navigation or for use in flight. It is simply provided as a tool. Use data at your own risk.*

Fee or Free: Free.

AVIATION DIRECTORIES

E-Flight Center

http://e-flight.com

RATING
✈ ✈ ✈

BRIEFING:
Primarily flight simulation-related. A closer look, however, reveals good commercial and private aviation info.

At first glance, E-Flight might seem a tad scattered—there's a lot of stuff dedicated to news, reports, and articles. Commercial and private aviation news, in the form of features and current events, abounds too. But, before scanning your bookmark list, grab your mouse and stay awhile. The current events are worth reading; the latest and greatest aircraft are profiled; there's vast amounts of airport info; and yes, a pilot's discussion lounge. Under the "Sims" area you'll find excellent, in-depth reviews on simulator add-ons and related products (example screen shots give you a heads-up on product quality).

Step into the E-Flight Center with coffee in hand, a mouse on the pad, and flight simulator questions in mind. Have a nice flight.

Fee or Free: Free.

28   200 Best Aviation Web Sites

AVIATION DIRECTORIES

Cyberflight

http://www.access.digex.net/~ooblick/cyberflight.html

RATING

BRIEFING:
A Maryland-based, East Coast directory site that's in complete disarray. The redeeming quality? It's roll-out-of-your-chair funny.

Where to start describing Cyberflight... Um, it's an autobiographical diary of perilous cross-country expeditions and a hodgepodge of links (some even non-aviation related). Toss into the cyber-blender a little sarcasm mixed with rambling commentary, and you've got an amazingly entertaining online place to visit. Don't forget to check out the embedded photos—they're a riot as well.

Once you cut through numerous solicitations for "lots of money" (donations, I guess), you'll learn where to fly to food; find "cool" airports; find what's new at Cyberflight; scan aerobatics clubs; and stumble across a host of miscellaneous stuff.

Although there's actually some good info here, it's a great place to visit for some old-fashioned hysterical high jinks.

Fee or Free: Free.

AVIATION DIRECTORIES

R/C Web Directory
Master List of All Sites
http://www.towerhobbies.com/rcwmaster.html

RATING
✢ ✢

BRIEFING:
Not pretty, but you'll find pages and pages and pages of remote control aircraft links. Not sure where to find radio-controlled (r/c) info? Start here.

During your initial visit, you'll just have to trust me that every radio-controlled aircraft link is here. This "old-style" laundry list of links does have everything—it's just painful to sift through it all. Frequent visitors may want to rely on the "new" tags for the most recent additions.

Loosely organized by headings, you'll find r/c links to: aircraft info and files; airplane-oriented sites; sailplane-oriented sites; helicopter-oriented sites; r/c clubs and organizations (broken down by state/country); r/c manufacturers & suppliers; r/c magazines; and miscellaneous sites.

Have your favorite beverage and snack standing by, you'll be here awhile if you're into r/c stuff. It's just a mind-boggling, grocery list of r/c that would cause an enthusiast to hyperventilate.

FYI: Although mostly r/c aircraft, you'll need to sidestep a few boats, cars, trucks, and other inferior vehicle categories.

Fee or Free: Free.

AVIATION DIRECTORIES

Charlie Alpha's Home Page

http://www.hiway.co.uk/aviation/aviation.html

RATING
++

BRIEFING:
Bravo to Charlie Alpha for his diligence in promoting UK/Europe flying. It's quite simply the UK's home base home page.

I recommend a stopover here for jumbled, all-over-the-place, punctuation-less brilliance. Granted, its limited scope emphasizes general aviation in the UK. A few things, at the time of review, were disabled or under construction. But, sidestepping the landmines unearths some glimmering gems. For example, you'll find *Pilot* magazine's link here (mostly UK, but some articles pertain to all pilots); events; associations; reader requests for information (online forum); great conversion tables (demystifies pressure, speed, temperature, time, volume, weight, and others); weather info; G-registration look-ups; and lots of quality links.

If you are interested in UK/Europe flying, there's enough knowledge, resources, and perspectives here to burn the better part of a day.

Fee or Free: Free.

AVIATION DIRECTORIES

Aerolink

http://www.aerolink.com

RATING

BRIEFING:

An award-winning global info-source that overflows with clickable mastery.

Fashionably designed and desperately needed, Aerolink really shines with its well-oiled search engine machinery. Peer into its linked offerings, and you'll get goose bumps over its simple, yet practical efficiency.

Four menu options provide clickable doorways into this aviation "link central." The perfectly simple icons of AeroWhat?, Add a Link, Link Central, and Star Sites begin your journey. (First-timers may want to check into AeroWhat? before jumping into the Link Central list of hyperlinks.) Major search categories in which you'll find yourself pleasantly inundated include: air carriers & airports, commercial marketplace, general aviation sites, government/regulatory sites, military sites, noncommercial sites, rotary wing sites, and sites of current interest.

Next, step into the Star Sites for some of the Web's best aviation sites—as reviewed by AeroLink's administrators. Sites are chosen monthly and offer corresponding "diamond" rating designations. And, finally, for adventurous surfers who happen to have their own aviation-related home page, Add a Link does just what it says—easily.

I know. You're thinking no site can be this good. Well, just type in the address and thank me later.

Fee or Free: Free.

AVIATION DIRECTORIES

Aero-Web
Aviation Enthusiast's Corner
http://www.brooklyn.cuny.edu/rec/air

RATING

BRIEFING:

Award-winning aviation-related features. Stop by for content—it's worth the ride.

Meander through the contents and you'll uncover this site's reason for being: variety. Although it's pleasant graphically, you'll really appreciate the nice smattering of useful information. Maintained by the hardworking volunteers at Brooklyn College, categories include: a museum index by location; an aircraft locator by type and manufacturer (here you'll find a long list of aircraft with some combination of descriptions, performance, specs, and museum display location); an airshow location and performance index; airshows by month; aviation history features; aviation records, some local New York stuff; and more.

Even while skipping among Aero-Web's offerings, you'll never lose your place—thanks to handy page menu headers. So, let loose and have some fun here. True enthusiasts will learn something.

Fee or Free: Free.

World of Flight (WOF)

http://www.wmich.edu/aviation/wof

RATING

BRIEFING:

An efficiently organized concentration of pilot tools and resources. It's sort of a one-stop-shop for great aviation info freebies.

Winning global aviation approval, World of Flight takes the Polar Route straight to cool offerings. Developed and maintained by an aviation sciences student at Michigan University, WOF makes pleasant work out of searching for vital aviation info. The graphic presentation is excellent *and* efficient—a powerful Web combo. In addition to the colorful main menu, subsequent pages also contain the same clickable icons. So, navigating throughout the site is worry-free and fun. You'll find popular general aviation aircraft pictures under Aircraft Images & Info; The Flight Deck Directory peeks into favorite cockpit layouts; and the Tutorials Section describes and informs with a focus on important aeronautical topics. The Weather icon opens into free, up-to-date radar and satellite images; Online Aviation Utilities provide links to necessary pilot searches (including N-number, airport, identifiers, and navigation aids); and Aviation Organizations compiles a long list of the industry's favorite associations.

Some extra time on your hands? I invite you to tap into: Software, What's New, Aircraft Manufacturers, Cool Sites, and Flight Simulation.

Fee or Free: Free.

Women in Aviation
Resource Center

http://www.aircruise.com/aca/wia

RATING
✈ ✈ ✈

BRIEFING:
This female-only forum opens the door into a world of women-in-aviation info.

With similar veracity as Amelia Earhart, the Women in Aviation site sheds important light on resources that might have appeared dim. This site firmly establishes an educational link to seemingly hard-to-find resources, such as: books, education and training, mailing lists, museums, networking online, organizations, publications, upcoming events, video tapes, women-in-aviation in business, and more.

Cleanly arranged with quick access, this cyber-forum makes networking convenient. There's a list of women-in-aviation contacts, complete with titles and e-mail accounts. Or, participate in open discussions with the site's online forum. Whether you're here to browse or choose interactive involvement, straightforward directions and descriptions abound.

Fee or Free: Free.

AVIATION DIRECTORIES

Bookmarkable Listings

Alex's Helicopter Home Page
http://www.geocities.com/CapeCanaveral/3838
Model and full-size helicopter information and resources.

The Flight Deck
http://www.theflightdeck.com
An aviation information services directory (mostly California-based info).

Air Cargo Online
http://www.cargo-online.com
Central cargo database of available air charter capacity worldwide.

R/C Airplanes NET
http://www.rcairplanes.com
Specific areas of info concerning radio-controlled airplanes based upon user requests.

AVLINK
http://www.flightdata.com/avlink/index.html
Aviation-related news and worldwide links.

The Flying High Page
http://shoga.wwa.com/~teleman/flyinghigh.html
Personal Web site offering categorized links and miscellaneous Chicago-based info.

Air Cargo Newsgroup Home Page
http://www.mta-ic.com
A Web site hub for the Air Cargo Newsgroup: misc.transport.air-industry.cargo.

Seaox Air-Medical Page
http://www.seaox.com
Air medical links, graphics, and industry info.

Airlines on the Web
http://www.itn.net/airlines
Extensive directory of airline Web sites broken down by category.

The Pilot Pitstop
http://www.mindspring.com/~anna716/index.html
Miscellaneous, non-categorized list of aviation links.

AviationDirectory.com
http://www.aviationdirectory.com
Fee-oriented "yellow pages" of aviation-related businesses.

Isreali Airpark
http://www.airpark.org.il
Articles, information, and links for the worldwide aviation community.

AviationNet
http://www.aeps.com/aeps/avnethm.html
Industry resource center offering news, link directory, schools, and more.

Calin's Aviation Index
http://www.calinsai.com
A central source of aviation contacts and news.

Airshow.com
http://www.airshow.com
Collection of current airshow schedules, performers, and related information.

Airship and Blimp Resources
http://www.hotairship.com
A volunteer effort to provide airship information to newcomers and veteran aeronauts.

AVIATION DIRECTORIES

Aileron's Place
http://www.akula.com/~aileron
A personal collection of aviation and flight simulator links.

Antique Aircraft Enthusiasts
http://www.wingsofhistory.org
Link resource for antique aircraft, including museums, organizations, and directory sites.

Army Aviation Directory
http://www.Jiga-Watt.com/AVNDirectory
Voluntary database used to locate former Army aviation personnel.

Aviation Business Center
http://www.airsport.com
Links to businesses supplying products, services, and information to the aviation community.

Aviation Organizations & Associations

Center for Advanced Aviation System Development (CAASD)

http://www.caasd.org

RATING
++

BRIEFING:
An FAA-funded, not-for-profit organization researching important aviation topics.

Just when you thought this dry, governmentally funded yawn machine could put you easily to sleep, I urge you to shake off the drowsiness and look deeper into its fascinating core of hot topics and research areas. Their goal: "operating in the public interest to resolve global aviation issues through research." Wait. Before you move to your bookmark list, click through its five categories: about CAASD, reaching CAASD, accomplishments, research areas, and hot topics.

I found myself leaning more toward the research areas and hot topics. I think you'll discover some thorough, insightful information (including charts) relating to: air traffic control, traffic flow management communications, navigation (excellent!), surveillance, systems & network management, architecture, airports, Free Flight, user request evaluation tools, and automated surveillance.

The content will pleasantly surprise—just slip past the "plain Jane" shell.

Fee or Free: Free.

The Ninety-Nines
International Organization of Women Pilots

http://www.ninety-nines.org

RATING
‡ ‡

BRIEFING:

Cyber-headquarters for the women-only Ninety-Nines.

Originally funded in 1929 by ninety-nine licensed women pilots, The Ninety-Nines organization catapults its promotion of flight fellowship with this thoroughly informative online presence. Although its crude design and lackluster site navigation offer a minor downside, the benefits are well worth the visit.

A bookmarkable must for all female aviation enthusiasts, The Ninety-Nines pages provide unending resources. In addition to organization and membership info, other topics include: The Ninety-Nines in Aviation History, Women in Aviation History, Forest of Friendship, 1929 Invitation, Women Pilots Today, President's Perspective, Women Pilot Exhibit, Learn to Fly, Future Women Pilot Program, Scholarships, Grants & Awards, Calendar of Events, Aerospace Education, Air Races, Airmarking, and more.

Member communication is conveniently available through an established private forum on CompuServe. See site for details on access.

Fee or Free: Free unless you join.

Women in Aviation International (WAI)

http://www.wiai.org

RATING
+++

BRIEFING:
Expanding the lofty horizons of active women pilots and wanna-bes.

Piloting a national and international cyber-course, the Women in Aviation International pages hone in on needed resources for aviation's women enthusiasts. Female site-seekers may be disappointed at first with WAI's archaic navigation (get ready to use your browser's "back" button frequently). But, once you begin clicking into its content, you'll feel better.

Tapping into Events gives you dates, times, and places of WAI-specific events. WAI News offers topical press releases. Education & Scholarships gets into the good stuff with profitable scholarship info. Your Career takes you to a current aviation job list with descriptions (a tad sparse at review time). The Archives area provides a semi-entertaining, pictorial WIA scrapbook without text. And, yes, there's Online Shopping available for WAI necessities. President's Welcome, membership info, conference dates, *WAI Magazine* (cover photos only), guestbook, and Web links are also at your clicking fingertips.

With their self-proclamation, "Women in Aviation is dedicated to the encouragement and advancement of women in all aviation career fields of interests," WAI and its site are great first steps for aviation-interested ladies.

Fee or Free: Free to view, fee-oriented membership.

The National Transportation Safety Board (NTSB)

http://www.ntsb.gov

RATING
+++

BRIEFING:
An online look into the darker side of aviation—the accidents.

Not a lot of fluff and graphical gadgetry take flight here. Obviously, subjects with which the NTSB deals don't lend themselves to frivolity. Uncover the site's dry contents and you'll realize why its investigation wisdom becomes instantly worthwhile. Why? Because learning from the mistakes of other aviators is relatively painless.

If you're curious (although I can't imagine why), you'll find yawning tidbits of miscellany, like: about the NTSB (history, mission, board members, organization charts, etc.); upcoming events (board meetings, hearings, and more); speeches and testimony by board members and staff; press releases; and more.

Hopefully, though, your real investigative interests will guide you to the insights and descriptions of over 37,000 aviation accidents—there's even a handy database search capability! Or, clicking into aviation accident statistics reveals tables from the annual aviation accident press release, passenger fatality accident tables, and most recent monthly statistics. And, when you're ready to learn more, lists of accident reports and studies may be ordered.

If an accident should happen to you, info on NTSB reporting is easily found online.

Fee or Free: Free.

National Aeronautics and Space Administration (NASA)

http://www.nasa.gov

RATING

BRIEFING:
Blast off into this bookmark-bound space voyage—it's quite a ride.

Building award-winning Web sites for aerospace surfers isn't rocket science. Then again, compared to the thousands of rejected sites I've hastily passed up, maybe it is.

As if the online mission control team were holding your hand through each page and topic, site navigation is effortless. Brilliant description-oriented links combine with fancy quick-click icons to launch you into unearthly euphoria. Uncluttered page layouts give you fascinating info, helpful pics, and plenty of easy-reading "white space."

Skipping directly into the Office of Aeronautics & Space Transportation Technology area, you'll climb into juicy research topics like: high speed travel, general aviation revitalization, next generation design tools & experimental aircraft, current events, library, commercial technology, and more. For other NASA favorites click into: Today@NASA (breaking news and project details); Q&A (great collection of thought-provoking questions and thorough answers); Gallery (video, audio, and still images); and Space Science (planetary exploration, astronomy, & research into the origins of life).

Fee or Free: Free.

Career Pilots Association (CPA)

http://www.erols.com/burnside/cpa.htm

RATING
++

BRIEFING:
Quick, clean, info-rich career pilot stuff. And, yes, they do lean toward membership.

The CPA page? A breath of non-graphical fresh air. If you're a career pilot type, give your modem a break and breeze through valuable news, insight, and membership info. Subjects conveniently at the ready include: Learn More About CPA; Meet Our Advisory Board; Read *Working Pilot,* CPA's Newsletter; Recent Press Releases; Position Papers; Membership Services (there's a bunch); Membership Types; Membership Application (simple type-and-submit form); Congressional/FAA Policy Documents (interesting items about pilot record-keeping, FAA notice on age 60 Rule, and single-engine IFR Part 135 operations); and Aviation-Related links.

Yes, you'll find some degree of aviation links on most sites. Hey, it's value-added for your browsing time. CPA's no different. Soon to be a searchable index, CPA's links go on forever. Categories include: FAA, NASA & NOAA sites, airlines, airports, associations, colleges & universities, FBOs & flight schools, helicopter sites, aviation museums, aviation images, aviation companies & commercial sites, other aviation site lists, and miscellaneous aviation resources.

Fee or Free: Free unless you become a member.

Air Force Link

http://www.af.mil

RATING

BRIEFING:
Official U.S. Air Force site proudly serves its online viewers with regimented excellence.

A model of Web perfection, the Air Force Link easily busts through my scale's five-plane ceiling. Criteria I've set aside for site organization, content, graphics, and intended audience are masterfully adhered to with style and purpose.

The opening menu gives you a hint as to the Web wonders you'll find lurking inside. Loading speed and appropriate presentation are perfectly balanced. But, before you get too dazzled, slip into the complete site map navigation page—arranged in graphical "tree" format or text-only map. Going a step further into navigational bliss, a site searching tool is also located on the page. With the Air Force Link's resources, you just can't get lost.

Content-wise, a fantastic array of goodies begins with News, Careers, Library, Sites, Images, Spotlight & Top Story. Specifically, Air Force Career subcategories include: civilian, enlisted, officer, and retiree—all come complete with pay charts. And, the Images page serves up an enormous list of high-resolution pics—a search tool just for photos makes browsing easy.

The Air Force Link successfully defends its surfers against the ghastly mess so often found in aviation Web offerings. Click in and see for yourself.

Fee or Free: Free.

The NATCA Voice

http://www.natcavoice.org

RATING

BRIEFING:
National Air Traffic Controllers Association members have a bit more to say than just, "traffic at your eleven o'clock position."

Curious about the voice behind that terse air traffic controller instruction? Get a better fix on that scratchy radio enigma with the cyber-version of The NATCA Voice. An info-packed forum for the National Air Traffic Controllers Association (NATCA), The Voice shouts its political opinions and speaks to important aviation legislation issues.

This no-nonsense forum brings you well-written articles and hard-hitting insight into the NATCA. Although site navigation isn't quite as good as their flight following, you will find a somewhat helpful site map, along with omnipresent main menus. Once you do delve into a topic of choice, you'll exchange the cloud deck of confusion with good information and politically charged commentary.

The main page takes you quickly into "current events that need your attention." After that you're invited to tap into: The NATCA Voice Shop (items to support your "voice" and union), The NATCA Voice Archive (archived newsletter articles), What's New, Main Areas (NATCA, aviation, labor, and legislative), Web Links, The NATCA Voice BBS, and Site Map. Looking for something particular? Just search the site. Hey, with the folks at the National Air Traffic Controllers Association, you'll easily find your topic destination. It's what they do.

Fee or Free: Free.

Cessna Pilots Association (CPA)

http://www.cessna.org

RATING
+ + +

BRIEFING:
Compilation of Cessna-specific, technical insight mainly geared to the CPA member.

It's no surprise that one of the world's most popular flying machines has its own association. Better still, the Cessna Pilots Association now provides its members and nonmembers this outstanding online technical information center.

Mainly a gentle nudge into CPA membership, the site does a good job of describing each available Cessna-specific resource before you click. *CPA Magazine* talks about the monthly publication—devoted 100% to solid technical data about maintaining and operating Cessnas. Tech Notes and Handouts store a series of helpful documents ranging from FAQs concerning Cessna problems to sources of discount parts. Cessna owners will also appreciate the Technical Hot-Line, Technical Library, Cessna Buyer's Guides, Systems & Procedures Courses, and Group Aircraft Insurance Program.

Updates are handled regularly with the promise of a future expansion to include a members-only area with: an online version of the latest issue of *CPA Magazine*, access to CPA's entire library of Tech Notes, and online discussions of Cessna-related issues.

Fee or Free: Free, but the good stuff requires membership in CPA.

Of course, as with all organized organizations, online membership is clickably convenient. If you're a Cessna owner, it's a "no-brainer."

International Miniature Aircraft Association (IMAA)

http://www.fly-imaa.org

RATING
✈✈

BRIEFING:
Large scale, radio-controlled cyber-center for fellow enthusiasts.

You can almost hear that high pitch engine whine before you reach the IMAA's pages. Those in the know, however, have become familiar with the more beefy sounds emanating from these larger-than-life, radio-controlled beauties. The next time you look up and see a Red Baron Super Stearman, a Byron Staggerwing, or a P51D, its pilot might just be on the ground with you.

Nourishing the large scale radio-controlled craze, The IMAA offers its online "hub" as a great central info resource. Here you'll pick up the course for local listings of chapters by district, as well as individual district news and photo coverage. A fantastic site index quickly breaks down the goodies: How to Join IMAA, How to Sanction an IMAA Event ("Rally of the Giants" for example), *High Flight Magazine* (IMAA's official printed publication) and Article Archives, IMAA Events Calendar, Announcements, and Flight Gear Accessories.

If time is limited, do make a touch-and-go into the photo coverage and reports from the last Rally of the Giants. These cherished large scale aircraft are amazing to behold—even digitally. And, if you're an enthusiast don't miss IMAA classifieds. Take advantage of free personal classifieds, or scan through the commercial ad listings.

Fee or Free: Free unless you join IMAA.

The IMAA site. It's a huge miniature aircraft haven. Grab the controls and stay awhile.

Helicopter Association International (HAI)

http://www.rotor.com

RATING
+++

BRIEFING:
Your online chopper choice, courtesy of The Helicopter Association International.

Fee or Free: Free parts searching for all! Parts listing for HAI non-members is free for a limited time.

Non-fixed wing fans need only arrive at this cyber-pad and clear some room on the bookmark list. You'll be back again.

Although the HAI site's in kind of a design spiral, sprawling volumes of worthwhile content more than compensate. Neatly arranged in columnar, text-based hyperlinks, rotor surfers will revel in clickable delight. Heliport searching is simply a matter of entering a city, state, or heliport. The Events Calendar & Heli-expo info give you a heads up on shows near you. And, the HELicopter Parts Search (HELPS) database teams parts listers and parts searchers effortlessly, for free!

Packed into the HAI site is a content load that would make an Aerospatiale Lamb 315B buckle. Click for industry news and archived info updated daily, online publications, chat rooms, action alerts, bulletins, hot spots, education, conferences and jobs, e-mail list services, regulatory issues, aircraft for sale/lease, and way more than I have room to list.

Among my site favorites is the Cornerstone feature. This computerized civilian/military registry identifies standouts in the helicopter industry. Read through biographical data and scan pics of thousands of helicopter professionals. It's a great way to track down a friend or colleague.

200 Best Aviation Web Sites

AirLifeLine

http://www.airlifeline.org

RATING
++

BRIEFING:
Find the online cyber-scoop for those in and around medical missions.

Just as the World Wide Web offers up a 24-hour labyrinth of resources, AirLifeLine proudly stands by with its own important network. As outlined in its online presence, AirLifeLine is a "nonprofit charitable organization of private pilots who donate their time, skills, aircraft, and fuel to fly medical missions."

The site clearly identifies organization goals and a myriad of FAQs with a simple contents list. Take a quick jump into: Services We Offer, Who May Use AirLifeLine, How Can You Help, AirLifeLine Volunteer Pilots, and Mission Requests. You'll quickly notice AirLifeLine skips the fancy GIFs and graphical fanfare associated with most of today's Web sites. The obvious concentration leans toward organization info—both for potential pilots and those needing medical transport. For example, private pilots may learn more about volunteering by clicking into AirLifeLine and the Private Pilot. Here, you'll browse careful text-based details about the missions—who qualifies, how they are requested, and how each is accepted. Pilot's Liability, Requirements for Joining, and Membership Application Details are also at the ready.

Curious about signing up? Read the letters to AirLifeLine. You simply can't help but be inspired to fire up your trusty flying machine and make a difference.

Fee or Free: Free.

Fly-In.org

http://www.fly-in.org

RATING
✢✢✢✢

BRIEFING:
Oshkosh online—officially brought to you by the Experimental Aircraft Association (EAA).

Wittman Regional Airport in Oshkosh, Wisconsin. Perhaps you've heard of it. Every year a few aviation enthusiasts stop by for a week and mingle with airplanes and fellow aviators. Actually, the event numbers speak more clearly to its worldwide appeal: over 800,000 attendees, 12,000 airplanes, 2,800 show aircraft, 700 exhibitors, and 500 forums. Numbers like these require hordes of promotion. Enter Fly-In.org.

Peek through this virtual window of Fly-In.org to get the complete skinny on this year's Experimental Aircraft Association Oshkosh Convention. Although most of the site info during my review alluded to the upcoming event at the time, you'll get the same scoop for the next show. When's the next show? Just check the convenient countdown expressed in months and days. Then push on through the multicolored buttons that light your way into: Oshkosh Current Event Info, Oshkosh Shows, and Message from the EAA President.

Probably during your visit, you'll spend the most time boning up on the upcoming Oshkosh info. An organized display prompts you easily through the photos and descriptions of: Oshkosh arrival procedures, tentative schedule of performers, exhibitors, day-by-day info, where to stay, transportation, monthly updates, forums, and admission rates.

Fee or Free: Free.

American Institute of Aeronautics and Astronautics (AIAA)

http://www.aiaa.org

RATING
✛ ✛ ✛ ✛

BRIEFING:
The AIAA creates the epitome of Web excellence for aerospace resources.

Fee or Free: Free.

Consistent with its professional nature and world-recognized resources, the AIAA blasts off into cyberspace during its 65th year of excellence. Technically touching on aerospace support services, the AIAA site closes in upon the perfect online template for resource sites. In every category I hold dear (content, layout, functionality, and audience), this aviation presence dominates.

First, join me in a functionality analysis. The introductory page loads quickly and gives the aerospace enthusiasts the entire site summary at a glance. Click either the illustrated icon or the text link—in both cases you arrive at, and transition through, your destination frustration-free. Second, layout steals a page from some Web expert's manual. You'll wander through appropriately illustrated icons and perfectly proportioned text pages. Bandwidth-hogging pics are nonexistent and descriptions sparkle with clarity. Third, the overall intended audience not only includes AIAA members, but ALL enthusiasts involved in the arts, sciences, and technology of aerospace.

Finally, content should be reason enough for a bookmark. Stretch your mouse hand, and follow me into: career planning and placement services (mainly for members), AIAA Bulletin (industry news, services, events, employment services, and more), conferences & professional development, technical activities, publications & databases, and more.

World Flight 1997

http://www.worldflight.org

RATING

BRIEFING:
Soaring chronicles of the re-creation of Amelia Earhart's 1937 world flight.

At the time you're reading this review, World Flight 1997 has come and gone. The re-creation and completion of Amelia Earhart's 1937 World Flight is a thing of the past. My hope, though, is that this online time capsule will stick around awhile. The daily chronicles of Linda Finch's heroic flight are of the timeless variety.

Originally introduced as an interactive way for students to follow along with Ms. Finch's daily journey, World Flight 1997's site now becomes a cyber-shrine to this fantastic flight. Learn more with a look into crew bios, aviation facts (trip facts, photos, plane, engines, and history), flight map, and weather reports along the way. With limited time, you may want to jump directly to the pilot's log. This firsthand account of the journey is captivating. You'll find unedited honesty!

Mixed in with the interactive content and expertly chronicled journey, the site of World Flight 1997 soars triumphantly with sheer design as well. The intro page uses an illustrated cockpit to showcase the site's clickable contents. Continuing on to selected topics will also reveal clever illustrations and visual navigational aids.

Fee or Free: Free.

The Official Blue Angels Home Page

http://www.mdc.com/version2/blueangels/ba1.htm

RATING

BRIEFING:
A dizzying aerobatic spectacle in real life and online.

Blasting onto the scene in a spectacular blue streak, The McDonnell Douglas-sponsored Blue Angels page simply dazzles. With expected style and skill, this Navy promo team demonstrates Web prowess as well. The site's perfect blend of synchronized navigation, speed, and flashiness almost rivals that of their in-air shows. Layout, clickable menu graphics, and well-chosen photos begin the online adventure. Then, informative text tells the rest of the story. Get the initial briefing by clicking into the squadron history, covering aircraft types, missions, and objectives. Next, the look into personnel focuses on the professionals behind the glitz—demonstration pilots, C-130 pilots, support officers, and maintenance & support team. I also recommend a peek into the equipment too: the F/A-18 Hornet (complete with performance stats) and the supporting C-130 Hercules (affectionately known as Fat Albert Airlines).

By far my favorite site feature is the proud display of photos in the gallery. You'll be bombarded with maneuvers, jets, and formations—enough to keep you busy for awhile.

Fee or Free: Free.

USAF Thunderbirds

http://aeroweb.brooklyn.cuny.edu/events/perform/tb/tb.html

RATING
++

BRIEFING:
A fast, sleek, and thoroughly entertaining site almost mirroring the real-life stuff.

Also graceful in the online skies, the Thunderbirds display perfect pageantry and digitized delights. As you would expect, uniformity and solid organization provide the glue to hold it all together.

A bit light on photos, the site focuses more on history, technique, and people behind the scenes. The well-written descriptions offer a fresh perspective and more insight anyway. Fascinating Thunderbird info can be quickly found by clicking into these articles: The Thunderbird Legend Lives On, Thunderbird History, The F-16 Fighting Falcon, Pratt & Whitney F100-PW-220 Turbofan Engine, Pilots Display American Airpower, Maintainers Keep Thunderbirds Airborne, Support Critical Piece of Thunderbird Puzzle, and Quotes.

As you'd expect in this promo site, the current airshow schedule is conveniently clickable. Courtesy of the Aviation Enthusiast Corner (also an award-winning site in this book), you'll be hyperlinked to a long list of shows—complete with date, show title, and city. Further links provide actual location, contact, and performers.

Fee or Free: Free.

National Business Aircraft Association, Inc. (NBAA)

http://www.nbaa.org

RATING

BRIEFING:
An organized organization with a resourceful site committed to NBAA membership.

Almost needing a suit and tie to view, NBAA's site exudes professionalism, style, and a hard sell toward membership. The nonmember area offers inexhaustible info on products and services. Clickable subjects include: Welcome & What's Hot; NBAA Membership (get the sign-up details here); NBAA Seminars and Conventions (dates & times); NBAA Publications Online; Political Issues (info on aviation policy at the state and congressional levels); Business Aviation Products & Services; The NBAA History; and Travel Sense (business travel productivity software).

Existing NBAA member? Just fill out the application form and move into the member-only area.

Fee or Free: Some info is free for nonmembers. Join the NBAA and get into more good stuff.

Aviation Safety Connection (ASC)

http://www.aviation.org

RATING

BRIEFING:

Nonprofit site furthering air safety through discussion groups.

With an honest, thorough approach, the ASC jumps into the cockpit with all of us and guides us through important topics like: cockpit command responsibilities, human error factors, and behavior patterns. There's a pilot lounge, bulletin board, ready room discussion forum, a library, and review of accident reports.

The emphasis here is on fundamental safety issues. The ASC uses accident and incident case histories to fine-tune the decision-making process. Through its online offering, the ASC makes it easy for all of us to learn from the mistakes of others. Conveniently, message and reply forms are strategically placed throughout the site. Forms are available for: The Webmaster, messages, subscriptions (free registration to *Cockpit Leadership*); e-mail to the editor, discussion group area; and the librarian.

Quite simply, ASC makes it easy to be a safer pilot.

Fee or Free: Free.

Aircraft Owners & Pilots Association (AOPA)

http://www.aopa.org

RATING

BRIEFING:
Yes. It's a recruiting tool. But, if you're interested in joining the world's largest general aviation advocacy organization, you can sign up here.

From magazines to trade shows and fly-ins to their Air Safety Foundation, AOPA isn't used to taking the number two position. Just like everything AOPA does, this site included, skillful organization has propelled this necessary association into the aviation limelight since 1939. This site, developed for membership and club benefits info, makes it easy to see why AOPA ranks among the 100 largest membership organizations in the United States. You'll find current fly-in info, pilot news, an intro into learning to fly, a sample of their printed magazine (*AOPA Pilot*), and yes, membership information (fees, application, etc.).

Existing members may tap into their own section for: 24-hour access to searchable databases, weather info, back issues of *AOPA Pilot*, events, and more.

Fee or Free: Free to view, fee for AOPA membership.

200 Best Aviation Web Sites   **59**

National Aeronautic Association (NAA)

http://www.naa.ycg.org

RATING

BRIEFING:
Yet another worthwhile site promoting a worthwhile club. Have a look—it may be for you.

Looking to entice new members into aviation and air sports, NAA's site invites you to grab your GPS and explore a "world of aviation" through NAA membership. Sign-up info, mission statement, corporate membership, affiliates, contacts, and a look at "NAA Today" are all here. Fun little tidbits include: aeronautical records, aero clubs, and air sports associations.

Although some weren't available during review time, special links to air sports include: Academy of Model Aeronautics, Balloon Federation of America, Experimental Aircraft Association, Helicopter Club of America, International Aerobatics Club, Soaring Society of America, United States Hang Gliding Association, and the United States Ultralight Association.

Fee or Free: Free to view, fees for NAA membership.

MicroWINGS

http://www.microwings.com

RATING

BRIEFING:

Informative online hangout for flight simulator types.

Yes, I've even uncovered an association for flight simulator buffs. Here's a reliable, helpful buddy to extract the maximum simulation exhilaration from your favorite flight software. As demonstrated by this site, MicroWINGS fiercely devotes itself to all types of aerospace simulation.

If you're searching for real, approved flight training or just a chance to dip into the fantasy slipstream, I've found your land-based cockpit pros. Although still under some construction as of review time, the site has a zillion products to review in the MicroWINGS Store, new product news, online chatting, and a strong push towards membership in the International Association for Aerospace Simulations. With membership, though, comes hefty product discounts, a full-color magazine subscription, free software, simulator bulletin board access, and more.

Fee or Free: Free to view, but if you're into simulators get the membership.

Experimental Aircraft Association (EAA)

http://www.eaa.org

RATING
++

BRIEFING:
Here's an Oshkosh junkie's fix for the whole year.

Wish the Oshkosh fly-in lasted year-round? Before and after the real thing, join EAA members at the official EAA home page to satiate your hunger for fun flying. Although still under construction as of review time, there's already plenty of well-organized info and plane pics in many clickable areas: History of EAA, EAA Aviation Foundation, EAA News, Young Eagles, Oshkosh Fly-In, EAA on TV, Ultimate Flights Show, Aviation Links, and EAA Membership Info.

Great site and organization for enthusiasts of every age and interest—"pilots, designers, builders, dreamers, and doers."

Fee or Free: Free to view, fee for membership.

International Aerobatic Club (IAC)

http://acro.harvard.edu/IAC/iac_homepg.html

RATING
++

BRIEFING:
Those having an aversion to staying upright will find no downside with this aeronautic club info.

So, you've had enough of straight and level, and you're looking to get inverted. Sift through the nicely organized pages of the IAC to learn more about aerobatics and membership with the IAC. To join, you'll need to be a member of the Experimental Aircraft Association (IAC is a division of the EAA). There's online application info; phone numbers and addresses for the main office in Oshkosh; and a long list of IAC chapters worldwide (some have Web pages).

As with any well-organized club, internal communications are a necessity. And, you'll find it's no different here. At your disposal are lists for an e-mail distribution and IAC snail mail addresses. (By the way, you don't need to be a member to be on IAC's address list.) Member or not, you'll find aerobatic info galore: general aviation servers, aviation databases, aviation reports & weather, other aeronautic associations, competitions, soaring, U.S. Aerobatic Team news, IAC rule changes, list of judges, IAC news (recently updated), as well as a few links.

It's enough info to make your head (and your aircraft) spin.

Fee or Free: Free to view, fee to join IAC.

Soaring Society of America (SSA)

http://acro.harvard.edu/SSA/ssa_homepg.html

RATING
✈✈✈

BRIEFING:
An online meeting place for Soaring Society members as well as nonmembers.

You've seen them in their tiny bullet-shaped cockpits floating on elongated wings. For those that glide, the thrill of soaring is infectious. The remedy is to join other thermal-hungry friends online. The official pages of the Soaring Society of America are found here—giving you a link to all phases of gliding nationally, as well as internationally.

The info contained here is organized nicely, graciously avoiding gratuitous photos. You'll find: a member resource book, safety-related articles, contest results; SAA Club details, news, and bylaws; list of U.S. soaring sites; calendar of events (official and unofficial); soaring magazine articles; e-mail address list; international soaring organizations; and affiliates/divisions.

Soaring buffs and wanna-bes should look into membership. Judging by this site's organization, you'll be in good hands.

Fee or Free: Free to view, fee for membership.

Federal Aviation Administration (FAA)

http://www.faa.gov

RATING

BRIEFING:
This information-rich governmental site will dazzle you with efficiency.

All government jokes aside (too many to list here), one can only stare slack-jawed at this wondrously efficient, visually appealing FAA piece of mastery. When the weight of federal bureaucracies begins to creep into your business/professional life, you'll find the light at tunnel's end here.

Granted, there's some administration stuff here that most will care less about. But, slip past it, using a nicely designed "Quick Jump" search engine to breeze through traditional red tape. You'll get your questions answered.

Or, examine some newsworthy current events like: GPS Transition Plan and Free Flight. There's FAA news and info, centers and regions, important FAA-related sites, and aviation safety info.

While some would call this photo-free site a tad dry, I see it as an itsy-bitsy, microscopic baby step toward governmental efficiency.

Fee or Free: Free.

Bureau of Transportation Statistics (BTS)
Office of Airline Information

http://www.bts.gov/oai

RATING
+++

BRIEFING:
An award-winning governmental (yes, governmental) site from the Bureau of Transportation Statistics that's shamefully useful.

Yet again, rather than rearing its ugly head of red tape, your government has chosen a course of surprising functionality and automation. Congratulations to the BTS for assembling this amazingly useful treasure of vital transportation information.

Tap into the BTS's Office of Airline Information and you'll open the statistical file cabinets housing: the FAA Statistical Handbook of Aviation, BTS Transportation Indicators, U.S./International Air Passenger and Freight Stats, and the Sources of Air Carrier Aviation Data. Although some stats may be a few years old, there's a wealth of info here on: airline on-time reports, passenger and freight counts, aircraft accidents, airport activity, general aviation aircraft info, aeronautical production, U.S. civil airmen, and more.

Fee or Free: Free.

World Aeronautics Association (WAA)
http://meer.net/users/waa/waaintro.htm

RATING
+ + +

BRIEFING:
Although the pages may not "wow" you, WAA's record-breaker stuff will.

Fee or Free: Free. All areas are accessible, except "Experimental Test Pilot's Club." Internet memberships are free (you'll get an e-mail newsletter subscription), and associate memberships are available for a nominal fee.

After wading through an otherwise long-winded history and purpose preface, you record-breaker types will get to the heart of the hype. Apparently encouraged by Zeus (I'm not sure why), those interested in aerospace records will find an uplifting online haven here. Of course before you reach the clickable record info, you'll be briefed on WAA membership and association benefits. But, once you wiggle past the opening sales pitch, things get a little more interesting. Pilots with the "right stuff" will get itchy just looking through the Categories and Record Areas (wingspan, speed, glider, fuel, propulsion, weight classes, aircraft categories, altitude, etc.). Also interesting are the WAA World Record Repository, The World Aeronautics Hall of Fame, and the Experimental Test Pilot's Club (requires membership—still under construction at review time). Find out how to set a world record and tap into the forms to document a flight.

While mostly text-based, the pages won't knock your socks off, but eager enthusiasts will find fun just the same.

International Wheelchair Aviators (IWA)

http://www.dsg.cs.tcd.ie

RATING

BRIEFING:
An able-bodied helping hand for an association of disabled, but inspiring aviators.

If you think for a minute that the heart of an aviator can be stifled by disability, point your mouse in the direction of the International Wheelchair Aviator home page. Although a little thin on beyond-U.S. stuff as of review time, the site explodes with pages of stories, inspiration, tips, resources, fly-ins, special ability flight schools, products, and more. When you scan through these pages, plan to spend some time— it's easy to get sidetracked navigating the embedded links.

Yes, you'll find some disorganization, and a few typos, but the content far outweighs these subtle distractions. Especially useful for pilots new to disabled flying, there are a series of resourceful Q&As. Answered questions include: Can a person who has a leg handicap, paraplegic, or amputee fly?; Do you need a special instructor to train you?; What kind of airplane can I fly?; Hand controls . . .?; Flight schools . . .?; and more.

The site has membership info and IWA benefits. If you join, you'll be in good international company. Current members originate from the U.S., Great Britain, Ireland, South Africa, Canada, Finland, Sweden, France, Germany, Portugal, and Australia.

Fee or Free: Free to view, nominal fee to join IWA.

Air Transport Association (ATA)

http://www.air-transport.org

RATING

BRIEFING:
When not embroiled in fare wars and cutthroat competition, your favorite commercial carriers come here to join forces, data, and knowledge.

Leveled off and cruising at commercial flight levels, you'll find the ATA's official Web site. This cyber-resource is always on time with 24-hours of ATA info for members, and a smattering of info gems for non-ATA types. Mainly serving its 24 commercial carriers (USAirways, United, Northwest, Southwest, TWA, etc.), ATA's well-organized site features an absence of unnecessary pics and slow-to-load graphics. A slick opening menu gets you into areas concerning general info, services, and member-only stuff. Don't shy away if you're a nonmember. There's a series of interesting things here for you too. The Airline Handbook launches into the history of aviation, deregulation, airline economics, how aircraft fly, the future of aviation, and more. You'll find industry stats, ATA member stock quotes, calendar (events and shows), an online aviation dictionary (under construction at review time).

Members can tap into an ATA Calendar; Airworthiness Directives; Memos and Data from the ATA's Engineering Maintenance and Materiel Council; Technical Support System; and the Prior Year Annual Report.

While we all have opinions on the efficiency of commercial carriers, you'll find ATA's site timely, useful, and convenient.

Fee or Free: Free with restricted membership area. See ATA's pages for details.

Angel Flight

http://xymox.palo-alto.ca.us/av/angelflight.html

RATING
+ + +

BRIEFING:
A nonprofit organization that not only moves you, but it transports those less fortunate with medical problems to a treatment destination.

Don't expect fancy graphics, gratuitous plane pics or rambling commentary here. The text-only descriptions and Q&A are quite enough to send your hope for the human race soaring. The Angel Flight site promotes and explains the aviation community's volunteer service of getting needy folks to diagnosis or treatment. Angel Flight's association of volunteer pilots and non-pilots join forces to: shuttle cancer patients to chemotherapy and surgery; carry people with kidney problems to obtain dialysis or transplants; and bring those with heart-related problems closer to treatment.

With this informative site, prospective volunteers will find many answers to common membership questions, including: Who belongs to Angel Flight?; Who does Angel Flight transport?; Where do calls come from?; Who pays for the flights?; How do I join?; What happens after I join?; and, What is my liability?

Angel Flight's pages are simply a persuasive call to action. Where do I sign?

Fee or Free: Free.

The Mechanic Home Page

http://www.the-mechanic.com

RATING

BRIEFING:
A quality resource for the true aviation maintenance pro—breathtakingly bookmarkable.

If you're of the variety who tinkers in the cowling or wrenches on undercarriage, wipe off the grease and reach for a mouse. Settling into The Mechanic Home Page reveals a heart-pounding depository of aircraft maintenance technician info. At the center of this site is the Aircraft Mechanics Fraternal Association (A.M.F.A.)—a craft-oriented, independent aviation union. Cleanly organized with great mechanic info, you'll find everything here is easily accessible. From several linked info pages to downloadable files to a few clickable icons, every site tool is at the ready. Sift through airline news (broken down by major carrier acronyms); enter or read comments on the bulletin board; read about the A.M.F.A.; download many important FAA files; scan the news archive; read observations from industry pros; tap into employment opportunities; learn from miscellaneous mechanic articles; and send e-mail to many industry-related mail boxes. The functional design and excellent variety of downloadable resources complete the package of perfection.

If you're remotely involved in the nuts and bolts side of aviation, break out your bookmark.

Fee or Free: Free.

Bookmarkable Listings

The World League of Air Traffic Controllers
http://www.geocities.com/CapeCanaveral/1140
The World League of Air Traffic Controllers site for news, chat, and contact info.

469th Security Forces Home Page
http://www.usafe.af.mil/bases/rhein/sp/spf.htm
Extensive look at the U.S. Air Force's 469th Security Forces assignments in Germany.

Vietnam Helicopter Flight Crew Network
http://www.vhfcn.org
A forum for recreational communications among aircrew members who served in Vietnam.

Naval Helicopter Association (NHA)
http://www.nasni.navy.mil/nha
Features NHA's *Rotor Review* magazine, as well as Navy, Marine Corp, and Coast Guard helicopters.

Lindbergh Foundation
http://www.mtn.org/lindfdtn
Informational source concerning the Charles A. and Anne Morrow Lindbergh Foundation.

United States Parachute Association (USPA)
http://www.uspa.org
Serving the only national skydiving association, the USPA, composed of over 33,000 members.

American Bonanza Society
http://www.bonanza.org
Informational site dedicated to the owners of Beechcraft Bonanza, Baron, and TravelAir aircraft.

Mooney Aircraft Pilots Association
http://www.311wc.com/mapa
Site helping to promote education in flying, operating, and maintaining Mooney aircraft.

United States Air Tour Association (USATA)
http://www.usata.com
Association news and information, current issues, proposed rules, and list of USATA operators.

Air Force Association (AFA)
http://www.afa.org
Nonprofit civilian organization promoting the importance of Air Force resources.

Institute of Navigation (ION)
http://www.ion.org
Informational site promoting the advancement of the art and science of navigation.

Civil Air Patrol
http://www.cap.af.mil
Online services and membership center for the Civil Air Patrol.

Airport Net
http://www.airport.org
Member resources for the American Association of Airport Executives and the International Association of Airport Executives.

Weather

The Weather Underground

http://www.wunderground.com

RATING

BRIEFING:
It's not a complete aviation weather provider, but it's sure a simple synopsis.

Most weather Webmasters tucked away in a dark cubicle somewhere have decided that you'll happily wait for high-byte graphics and giant radar maps to load. I, however, fall into the unhappy category when superfluous graphics and unwanted miscellany rain upon my weather inquiries.

If you're impatient too, go underground in the Weather Underground. It's amazingly easy on your modem and efficiently organized. Don't believe me? The first screen to load gives you three options to find your weather: 1) type your city and state into a search window; 2) click anywhere on the U.S. map; or 3) find your state's hyperlink and click. It's so fast, I become giddy with weather euphoria.

Once you've arrived at the desired city, a table of current conditions provides: time of report, temperature, humidity, wind, pressure, conditions, sunrise, sunset, and moon phase. Forecasts include descriptions and temperature for your desired city, as well as your state's extended forecast.

Fee or Free: Free.

USA Today—Weather for Pilots

http://www.usatoday.com/weather/wpilots0.htm

RATING
++

BRIEFING:
The big-name Web source that conveniently narrows its weather focus for pilots.

It's really your call whether or not you'd like to scan the online news from USA Today. However, I'd like to point you toward their specific weather resource for pilots, aptly named "Weather for Pilots." Mainly it's a collection of links specific to pilots and weather-related issues. The page does an excellent job describing the appropriate resources and offering a corresponding hyperlink. Unlike most online news sources of this caliber, graphics, pics, and charts are nonexistent. It's simply a quick way into the weather and related tidbits you need.

At the time of review, the site makes it easy to tap into: FAA Aviation Weather Research, Latest Info on Alaskan Volcanoes (important piloting info here on this busy route), NTSB Information Now on the Web, Flying Into Hurricanes, Pilots Report Hazards to NASA, NASA's Aviation Human Factors Research, and Online Weather Calculator (converting temperatures and calculating density altitude).

Perhaps a bit more specific to weather, the following topics catapult you into a great collection of associated links: thunderstorms, icing, live weather, ground school, and more.

Fee or Free: Free.

National Weather Service

http://www.nws.noaa.gov

RATING
✈ ✈ ✈

BRIEFING:
Weather service pros give you a 24-hour option for getting the real scoop on Mother Nature's intentions.

Partially obscured among the countless weather resources found on the National Weather Service's site, you'll find a healthy grouping of aviation products. Get current info from Terminal Aerodrome Forecasts, aviation weather discussions, aviation METAR reports, Terminal Forecasts, and more. Mostly text-based, the available forecasts and observations offer accuracy and speed. You won't be waiting for maps or graphics to load.

Pulling back from the focus on aviation, the National Weather Service site also thoroughly covers weather-related topics in a more general sense. When you've got some extra hangar time, be sure to scan through the Interactive Weather Information Network (warnings, zone, state, forecasts), black and white weather maps, U.S. weather bulletins, tropical cyclone warnings and products, fire weather, and Alaska products.

It may not necessarily be pretty, but all your weather is here from those that know.

Fee or Free: Free.

Aviation Weather Center

http://www.awc-kc.noaa.gov

RATING
+ + +

BRIEFING:
Preempt your local weather reporter with forecasts from the source.

Weather info doesn't get much closer to the source than this. Without pomp and pageantry, the National Weather Service and The National Oceanic & Atmospheric Administration serve up the Aviation Weather Center with an information-rich presentation. Be forewarned though, you won't come across colorful maps and pretty graphics. Make sure you're up to speed on coded weather info, and you'll breeze through its text-only forecast reports.

Up to 24,000 feet, U.S. forecasts mainly include warnings of flight hazards, such as turbulence, icing, low clouds, and reduced visibility. Above 24,000 feet the Aviation Weather Center provides warnings of wind sheer, thunderstorms, turbulence, icing and volcanic ash.

Site navigation is relatively easy with many "return to" links. While it may take awhile to become comfortable sifting through site contents, you'll eventually get to: AIRMETS, Area Aviation Forecasts, Domestic SIGMETS, Terminal Aerodrome Forecasts, TWEB Routes, Winds Aloft Forecasts, and more.

Fee or Free: Free.

WW2010—The Online Meteorology Guide

http://ww2010.atmos.uiuc.edu/(Gh)/guides/mtr/home.rxml

RATING

BRIEFING:
Meteorology introduction with a flair for current weather from your educated friends at the University of Illinois.

Always keeping a watchful eye on the weather and Web wonders, I'm always elated when I uncover a combination of the two as well done as WW2010—The Online Meteorology Guide. Sure you can get current weather here (handsome revision being made as of review time), but the real impact stems from a combo of current weather AND instructional modules.

Meteorology instructional modules delve into a variety of fascinating topics using charts, graphics, and easy-to-understand description. Light on Optics introduces how light interacts with atmospheric particles. Clouds & Precipitation introduces cloud classifications and developing precipitation. The Pressure Module defines pressure related to altitude. And, although this is just a taste of the additional contents, read on about: air masses, fronts, weather forecasting, severe storms, and hurricanes.

Thoughtful navigational features are almost too abundant to list. But it's worth noting some of my favorites: an option between full graphics or text-based site layout, a helper menu explaining the site navigation, color coded highlights for current location, and friendly left-margin menus throughout.

Fee or Free: Free.

Weather—Cable News Network Interactive (CNN)

http://www.cnn.com/WEATHER/index.html

RATING

BRIEFING:
Will the Web's weather wonders never cease? Get well-done weather, 24-hours, with CNN's interactive presence.

Adding to a swelling list of lofty weather sites, CNN's weather page blows onto the cyber-scene with its own worldly prognostication. As is often the case with super-sites of this caliber, visually appealing graphics and orchestrated presentation subtly make you feel warm and cozy.

Four main topic icons quickly move you into: U.S. Forecasts, World Forecasts, Weather Maps, or Storm Center. U.S. Forecasts, for example, uses simple text links and non-obtrusive visuals to give you fast weather data. First, choose from a region list: Northeast, South Central, Southwest, etc. Then, find your city. Up pops a corresponding four-day forecast, complete with highs/lows, precipitation, winds, pressure, and humidity. Curious about the bigger picture? Tap into your regional satellite and radar maps—they're easy to read and fast to load. Or, scan through 28 worldwide weather maps and images. From the Africa satellite image to the Europe forecast map, global weather's here for the clicking.

Although completely unrelated to aviation, a current news topic list follows you in the left margin. With a click you'll jump to CNN's current news on sports, travel, world, U.S., local, and more.

Fee or Free: Free.

American Weather Concepts (AWC)

http://www.amerwxcncpt.com

RATING

BRIEFING:
Functional, easily used, comprehensive data. Any more great features and this weather wonder could soon replace your local TV weather personality.

Functional design and great organization make AWC's weather site one of your first stops for comprehensive weather. Even if you're new to the online weather scene, the transition is easy with a graphical site explanation (explaining each section with maps). Once you're comfortable, have a look at: Subscribe, Help, What's New, Forecast Center, Doppler Radar, Live Wx, Aviation, Travel, Flyte Trax, Chit Chat, and Wx History.

Pay a nominal monthly fee for basic service (everything, but the individual NEXRAD Doppler Radar sites), or delve into full-blown weather euphoria with the enhanced service (an additional, yet still nominal monthly fee). The best part? No hurrying necessary with unlimited downloading!

Fee or Free: Fee-based. Tempt yourself with a 14-day free trial.

AccuWeather

RATING
✚ ✚ ✚ ✚ ✚

BRIEFING:
This award-winning weather site makes it easy to see why good design, layout, and organization speak volumes in online communication.

From the moment you enter the "lobby," there's no question that some graphics guru got ahold of these pages. Visual wizardry takes the form of icons, illustrations, and easy-to-see maps. But, more importantly, you'll enjoy future return trips due to careful organization. It's so easy to navigate with the masterful visual references. From the AccuWeather Lobby you'll be introduced to Get Weather—a sampling area of the more than 35,000 types of AccuData (AccuWeather's online database) information products. Free sample weather products have a time delay, but subscribers will receive real-time access to everything. AccuWeather's products include: satellite pics, NEXRAD Doppler Radar, weather maps—current forecasts, Ray-Ban UV Index, temperature band maps, current conditions, weather discussions, and more.

It's an organized look at weather, whether you subscribe or not.

Fee or Free: Free for some weather samples; fee-based for full-blown, real-time weather products.

Aviation Model Forecasts

http://wxp.eas.purdue.edu/aviation

RATING
✈ ✈ ✈

BRIEFING:
An educated predictor of aviation weather.

When talk among pilots turns to weather, things usually get more serious. Abandoning any hint of whimsical buffoonery, the Aviation Model Forecasts site sticks to the topic at hand without cracking so much as a smile. From the get-go you'll plunge into technical weather data and forecasting models. Here you'll tap into colorful contour plots for weather forecasts.

Originating from The Department of Earth and Atmospheric Sciences at Purdue University, the WXP Weather Processor analyzes and displays meteorological data and satellite images. Mostly designed for those on a meteorologist's level of understanding, this site does provide non-meteorologists the ability to view the data in varying degrees of complexity.

The plots are usually updated once every twelve hours and offer a full range of forecasting—from twelve hours to ten days. The index includes: individual plot summaries; general forecast plots; initial analyses; 12-, 24-, 36-, 48-, 60-, and 72-hour forecasts.

Yes it may take awhile to decipher, but you're looking at some highly reliable forecasting here.

Fee or Free: Free.

The Weather Channel

http://www.weather.com

RATING

BRIEFING:

The Weather Channel's masterful weather magicians mix visual delights with practicality.

The Weather Channel site—a great example of what happens when you mix talented designers, high-end software gadgetry, efficient hardware, and a team of marketing pros. Let this group loose on the cyber-world, and you get online brilliance. Although you may not get all the pieces of your aviation weather data here, you will get quick, accurate, searchable weather tidbits, including winds aloft.

If you're a weather enthusiast with a hankering for facts and trivia, you've found your utopia. Start by typing your selected state and city into the "Quick Weather Finder." Instantly you'll get stuff like: current temperature, wind speed/direction, relative humidity, five-day forecasts, barometric pressure, and current conditions.

After you've checked out all of your favorite cities, don your slickers and splash through the other fun items here. Dip into the Weather News, Weather Whys, All About Us, Weather and You, and even a few weather-related features. All of the clickable topics move you into logically presented information with visually captivating graphics—and the speed will pleasantly surprise you.

With this expert site, there's just no downside—unless the predicted weather keeps you grounded, that is.

Fee or Free: Free.

WeatherNet: WeatherSites

http://cirrus.sprl.umich.edu/wxnet/servers.html

RATING
✢ ✢ ✢

BRIEFING:
Whether you're ready or not for over 380 weather sites—it's here, it's unbelievable, and it's waiting for you.

Sure weather's important to all of us aviation types. But, what you've got here is weather obsession—in the most positive sense, that is. Luckily, the weather worshipping folks at WeatherNet have assembled easy access to over 380 North American weather sites. Yes, over 380. Get your bookmark list cleaned up and get ready to add. At review time the 380 sites are simply listed in alpha order (a tad unwieldy), but look for better organization coming soon.

Aside from this unprecedented library of weather links, you'll find a clickable gem of WeatherNet features. Highlights include: Forecasts and Warnings (organized by state), Radar and Satellite (cool, clickable U.S. map), Weather Cams (local photographic peeks into selected cities and popular resort destinations), and Travel Cities Weather.

So, if you've got a few minutes to dabble with Doppler or see through satellites, get out your umbrella—you'll always run into weather here.

Fee or Free: Free.

Weather by Intellicast

http://www.intellicast.com

RATING
✈ ✈ ✈

BRIEFING:
Knock-your-socks-off graphics combine with user-friendly organization to form weather magic.

You've heard the hype. Whether you've secretly embraced or shunned MS NBC's joint online ventures, you may find positive common ground in the form of Weather by Intellicast. Sure, it's just a sideshow for MS NBC News, but Intellicast has created a unique and well-organized look at worldwide weather. Marvelous graphics and useful visual delights reign supreme here.

There's hordes of cool maps, weather-related icons, and menus. Get started with a main menu that leads to: USA Weather, World Weather, Travelers, and Ski Reports. Searching by city is easy. There's a map of popular cities, as well as a clickable list to get to your favorite city's weather. Once you find your city of choice, you'll become weather savvy with instant info relating to temperatures, forecasts, and a host of images to view. The long list of weather images includes: radar, radar summary, satellite, NEXRAD, and precipitation.

While you're visiting, don't forget to check into the monthly almanac, special reports/features articles, and Ask Dr. Dewpoint. It's fun weather fancy for everyone.

Fee or Free: Free.

Bookmarkable Listings

Singer's Lock
http://www.weather.org
Online book offers a "twentieth century look" at meteorology.

The Weather Visualizer
http://covis.atmos.uiuc.edu/covis/visualizer
Online weather map with options for customizing your own maps and images.

Real-Time Weather Data
http://www.rap.ucar.edu/weather
Weather data categorized by satellite, radar, surface, upper-air, aviation, and more.

Charles Boley's Weather Stuff Online
http://www.cwbol.com
Personal collection of described links to current weather maps, radar images, and weather newsgroups.

Atmosphere Calculator
http://www.geocities.com/CapeCanaveral/1030/atmcalc.html
Calculates dewpoint, relative humidity, wind chill, heat index, and more based upon information given.

National Climatic Data Center (NCDC)
http://www.ncdc.noaa.gov
Online collection of resources from the NCDC—the world's largest active archive of weather data.

Sunrise/Sunset/Twilight and Moonrise/Moonset
http://tycho.usno.navy.mil/srss.html
Automatically calculates sunrise/sunset, twilight, moonrise/moonset with given longitude and latitude.

Pilot Resources

AirCharterNet

http://www.aircharternet.com

RATING
✈ ✈ ✈ ✈

BRIEFING:
Air charter made easy and almost automatic with a virtual charter agent.

Putting aside the fact that the animated propellers rotate awkwardly (slowing things down slightly), The AirCharterNet travel site reigns supreme in many cyber-areas: content, design, page navigation, and organization.

Air charter passengers, air charter operators, and frequent travelers in general need to make room on the bookmark list for this resourceful Web wonder. Once you've become a member (it's painless and free), you'll meet the efficient virtual charter agent who automatically takes care of everything. Without charging a nickel more for your flight, the virtual charter agent will identify qualified air charter operators for specific trips; swiftly and easily solicit price quotes from any one of them; e-mail price quotes to you; and help you reserve flights.

Beyond the vast resources of the virtual charter agent, you'll also find the site helpful for the travel-related topics of: vacation planning, ground services, worldwide weather, and filling empty charter legs.

Fee or Free:
Membership is free.

Fillup Flyer Fuel Finder

http://www.wdia.com/ff/ff

RATING
+ + +

BRIEFING:
Finding your way through this fuel finder is worthwhile.

Maybe someday Fillup Flyer Fuel Finder will find the time away from the pump to clean this confusing, yet bookmarkable online presence. Don't misunderstand, Fillup Flyer is in this book for a reason. Under the thin veil of design—odd multicolored buttons, clip art, and tailspin site navigation—lurks a very useable fuel resource that is worth the initial effort.

Fee-based, Fillup Flyer provides members and nonmembers fuel price reports based on routes, nonstop, multi-destination, area, or statewide. What are your options for report info delivery and requests? Most choose computer, but fax, voice, or mail are also available. Once you maneuver through a barrage of menus, submenus, "major level buttons," "sub-level buttons," and (my favorite) "twisting red and yellow arrows," you'll have the major topics in sight. Click into: About Fillup Flyer, Admin, Guest Log, Home Page, General Info Statistics, Membership and Costs, Member Reports, Nonmember Reports, Premier FBOs, Sample Reports, and more.

Hey, if a little extra scrolling and clicking is worth saving up to a dollar per gallon on your next trip, then tough it out and assign this one to your favorites list.

Fee or Free: Fee—sign up annually (recommended) or pay by report.

UK Airfields Online

http://www.uk-airfields.co.uk

RATING
✛ ✛ ✛

BRIEFING:
UK-only airfield directory dazzles with simple brilliance.

Okay, my aero Web search narrowed here just a bit. But, such a simple, yet functional gem deserves its moment in the limelight.

Handsomely designed, UK Airfields is aptly named with a singular purpose—giving you instant access to popular UK Airfield info. Site navigation is basic. Simply use the pull down menu, find your desired destination, and click "visit." Yes, that's it. There aren't any ads, banners, or membership requests. UK Airfields is around for your benefit—really.

As of review time, the list of possible airfields to visit includes: Belfast, Birmingham, Compton, Abbas, Duxford, Edinburgh, Exeter, Headcorn, Ipswich, Isle of Man, Lasham, Lundon Luton, Manchester, Norwich, and Wycombe Airpark. Once you make your choice, each airpark maintains its own pages. A sample selection? Belfast International Airport's site includes valuable info, like: getting there, Northern Ireland, business profile, airport facilities, flight schedules, flight arrivals, flight departures, and airport news.

Fee or Free: Free.

Aviation Information Resource Database

http://www.airbase1.com

RATING

BRIEFING:

Let your mouse do the clicking through this big yellow book of online aviation resources.

Cleared for an informative flyby, you are invited to mouse your way around this resourceful labyrinth aptly tagged the Aviation Information Resource Database. Brought to all cyber-flyers free of charge by AIRbase ONE, this dominating database equates to a computerized Yellow Pages.

Flip through its topics and you'll see what I mean. Delve into over 12,000 aviation businesses listed in over 1,100 categories. In addition to thousands of service-related listings, there's a complete facility directory of all public/private airports and heliports. Also at your curiosity's convenience are: aircraft and engine parts, FBOs, avgas or jet fuel suppliers, and an exhaustive general aviation events calendar.

When you're really ready to pinpoint a preference, eight searchable subjects get you there with powerful queries. Just pick a topic: aviation businesses, airports, fuel suppliers, fuel prices (updated daily!), lodging, restaurants, ground transportation, general aviation, events calendar, and more.

Although hard to do, getting lost at this airbase is corrected easily with a handy "Need Help?" button. It's a progressive taxi through an industrious airpark.

Fee or Free: Free.

FlightWatch

http://www.flightwatch.com

RATING
++

BRIEFING:

FlightWatch directs you to aviation's finest legal resources with your own online attorney.

The legalities of aviation? Although most of us are way off course, fluttering around helplessly in the complex airspace of law, may I propose a source of refreshment? Keep your eye on FlightWatch—Resources for Pilots and Aviation Lawyers.

It's not a visual wonder, but I prefer my legal resources less flashy anyway. Some of the bulleted topics offer descriptions before you link up, but most do not. Regardless, you'll encounter a nice variety of FAA documents, databases, and handy lookups. Some links tie into another of my favorite sites, AVWeb, for which you'll need a password.

Even if you're not in any legal trouble, being familiar with FlightWatch's facts, figures, and federal resources might just help you to avoid any forced hangar time. Clean up your bookmark list and make some room for this one. You never know when legal questions will arise for: airspace, airways, routes and reporting points, air taxi and commercial operations, aircraft registration, testimony, production of records, services of legal process, recording aircraft titles, security interests, and more.

My closing argument? It's a free, resourceful hub for aviation's legal questions. Case closed.

Fee or Free: Free.

PILOT RESOURCES

Equipped to Survive

http://www.equipped.com

RATING

BRIEFING:
The survival instinct is alive and well with this must-bookmark site.

Long flights over water. A perilous single-engine journey over mountainous terrain. What if disaster strikes? Will you be prepared? The fact of the matter is survival—you may have only one chance. Make it count with the online excellence of the Equipped to Survive site.

First a note on esthetics—don't expect any. Honestly proclaimed up front, the site's author emphasizes information, not imagery. Second, site navigation simply consists of links to topics and "previous page/next page" buttons. So, nothing fancy here either. However, dip into the third criteria of content and you strike gold. The focus, you'll find, is on equipment—what is useful, what works, and what doesn't. Most of the site info is based upon the author's research on wilderness and marine survival from an aviation perspective.

Fantastically insightful articles worth printing and saving include: The Survival Forum, Basic Aviation Survival Kit, Ditching (for pilots), Aviation Life Raft Reviews, Survival Skills and Techniques, Aviation Life Vest Reviews, and more.

Fee or Free: Free.

PILOT RESOURCES

TheTrip.com

http://www.thetrip.com

RATING

BRIEFING:
A travel agent, a taxi driver, a map interpreter, a hotel concierge, a maitre d', and an airport TelePrompTer all rolled up into an online info-fest.

Straying a tad from my aviation-only focus, I must include this favorite of mine for the frequent traveler. Whether you're the pilot in command, or some other MD-80 crew is getting you there, TheTrip.com becomes the ultimate in travel insurance.

Wonderfully educated design pros weave their obvious skills and combine perfect page scripting. Thoughtful navigational bars and subtle icons guide you effortlessly through fantastic travel data. Under The Flight, you'll become your own travel agent—checking flight availability and actually making reservations. There's even a real-time flight tracking query to check any flight's status. The Airport provides a host of interactive maps, guides, and transportation strategies. A thorough directory of hotel reviews and associated phone numbers are also a click away in The Hotel hyperlink. And, saving the best for last, The City category ably spews out specific weather, lodging, dining, or city map data based upon your chosen city.

Tired of business travel stress? Settle into TheTrip.com for a pre-departure briefing.

Fee or Free: Free.

96   *200 Best Aviation Web Sites*

PILOT RESOURCES

Air Safety Home Page

http://airsafe.com

RATING

BRIEFING:
Airline safety analysis gives you the hard truth and worthwhile advice for airline travel.

Odds are you were bombarded with media overload from the last airline disaster. Newspapers, radio, cable, and even Internet news join the bandwagon when heart-wrenching air carrier tragedies occur. In the interests of safety, this site's veteran airline safety analyst brings to light important industry info not easily found in one source.

Some events you may painfully remember, other airliner mishaps may have escaped your attention. In any case, the facts and observations found in the Air Safety Home Page may open your eyes to valuable passenger advice.

Complex navigational gizmos and graphics are virtually nonexistent. But, Air Safety's claim to acclaim is hard, reliable data and important advice. Read about the last ten fatal jet airliner mishaps, fatal jet airliner events by model, the top ten air traveler safety tips, child safety, top ten questions about airline safety, and more.

I know it may not be the most cheerful topic, but I urge you to point your browser here for safety's sake.

Fee or Free: Free.

PILOT RESOURCES

The Air Safety Investigation Resource (ASI)

http://startext.net/homes/mikem

RATING
✢ ✢ ✢ ✢

BRIEFING:
Strategic selection of quality aviation links with safety in mind.

Similar to my publication's goal of rounding up worthy sites, The Air Safety Investigation Resource seeks to narrow the infinite sea of aviation's online cache. Simply organized into a "catch-all" table of contents, the site topics dribble down the page.

Although you'll be keeping your scroll bar arrows and "back" button busy, this site's worth the extra hunting time. Navigation and aesthetics rank among the ho-hum variety, but the thoughtful collection of hyperlinks catapults it onto the bookmark list. Obviously the time spent sifting through and collecting quality reference sources took priority—and rightfully so.

Check into these grouped highlights: Incident/Accident Reports, Registration, Pilot, Airport, & Navigational Aid Databases; Regional & National Surface, Upper Air, and Pilot Advisory Maps; Celestial, Pharmaceutical, Aviation Medicine, Pesticide Online Reference Library, and Search Engines; Weather Observations, METARs, TAF, Zone Forecasts, Station Identifiers & Locations, and more.

Until an official online aviation library takes center stage in cyberspace, this is it.

Fee or Free: Free.

PILOT RESOURCES

Aircraft Technical Publishers (ATP)

http://www.atp.com

RATING
++

BRIEFING:
Constantly searching for Airworthiness Directives? Free up the modem and cruise through ATP's instant list!

While other reasons exist for visiting ATP's site, my bookmark list increased by one because they offer a thorough list of Airworthiness Directives (ADs). Even better, the list is continuously updated with all ADs issued within the last 30 days. The value for repair stations is obvious. It means fast, thorough research. Even air charter operators, brokers, and owners will benefit from its timely information.

At the heart of ATP's AD offerings is a quick interface. The somewhat confusing nature of ADs disappears with a summarized list of ADs less than 30 days old. The AD number, effective date, manufacturer, short description, and recurrence status are conveniently simplified into a clickable table. To actually view a chosen AD, you'll need the Adobe Acrobat Reader. Download it from the ATP site if you don't have it. Go on, it's free!

More marginal site attributes involve some maintenance products and services, *FLIGHTLine Newsletter* info archive (Adobe Acrobat Reader needed), employment opportunities, and more.

Fee or Free: Free.

PILOT RESOURCES

Aviation & Aerospace Medicine

http://www.ozemail.com.au/~dxw/avmed.html

RATING
++

BRIEFING:
A no-frills, one-stop shop of aerospace medicine knowledge.

It's not pretty. But topics like hypoxia, G-LOC, and the leans never are. The site's visual style seems to mirror the somewhat serious nature inherent in aviation medicine discussions. Don't expect any frivolous pics or illustrations. The focus remains solidly on aerospace medicine. Period.

Once you click in, you'll be astounded at the worldwide collection of articles and publications. The phenomenal scope and breadth spills into just about anything currently troubling you. Selections like: Alcohol & Aviation, The Senses During Flight, and Laser Corrective Eye Surgery for Pilots are just the beginning. Read on for a whole series of articles relating to hyperbatic medicine combined with foot ulcers, oxygen therapy, wound healing, and strokes.

Looking for a specific aerospace medicine contact? A convenient alpha search quickly locates any one of 170 practitioners in 27 countries. Or, you may be more interested in tracking down a specific aviation medicine organization. This site puts you in touch with groups like: The Aerospace Medicine Association, Safety and Flying Equipment Association, and The International Academy of Aviation & Space Medicine.

The medicinal resources are endless. The value here is obvious.

Fee or Free: Free.

PILOT RESOURCES

Great Circle Distance Calculator

http://www.atinet.org/~steve/cs150

RATING
✛ ✛

BRIEFING:

Dialing for distance? Call up this nifty 'Net calculator.

Born out of a software engineering project in 1994, the Great Circle Distance Calculator pinpoints distance faster than you can say "GPS." Obviously not meant to impress with grandiose design or layout, this global distance tool quietly does its job, perfectly. You'll crack an appreciative grin the moment you need a flight distance in a hurry.

The site simply computes the flight distance between any two points on earth, called the Great Circle Distance. As you may have guessed, the two points are generally derived from latitudes and longitudes (degrees, minutes, and seconds—or decimal format). However, if you're a little thin on "longs" and "lats," you're still covered. Just enter the desired three-letter airport codes. You'll get the same result. When shopping for more generic distances, try the clickable map. Though it's not as accurate, you'll probably get your global distance questions answered.

Whatever the input method, your resulting data may be viewed in statute miles, nautical miles, or kilometers. It's your choice, and it's free. How many online aviation tools can say that?

Fee or Free: Free.

PILOT RESOURCES

EarthCam

http://www.earthcam.com

RATING

BRIEFING:
Be everywhere at once with a 24-hour peek into the world's live video cams.

Room with a view. Window on the world. An eye in the sky. Whatever your description, EarthCam *is* the source. Blossoming into a global cyber-cam hub, EarthCam compiles and categorizes the limitless, living directory of live video cameras on the Web.

Generally used by aviation enthusiasts as a quick destination peek, these handy cams offer glimpses of live weather conditions—instantly. Though obviously not worthwhile for navigation or flight planning, the cams do give some quick insight into current conditions. Better still, the wide variety of cams isn't limited to airports or large cities. You'll get a look at the Bay Bridge traffic, or Chicago's lakefront, or Disney World.

Colorful page presentation, clever icons, and convenient searching tools cheerfully guide you into an unbelievable array of cams. Peer through traffic cams, weather cams, business cams, educational cams, scenic cams, and more. Or, get a bit more serious about weather with a clickable world map broken down into regional satellite views.

My favorite part? Previewing text descriptions before downloading the pics saves hours of frustration. Be sure to read them prior to clicking—there are some time-wasters in the mix.

Fee or Free: Free.

102 *200 Best Aviation Web Sites*

PILOT RESOURCES

The Air Charter Guide (ACG)

http://www.guides.com/acg

RATING

BRIEFING:
A must-be-viewed site for charter operators and passengers.

Whether you're on the passenger side or the cockpit side of air charter, you simply must have a printed and online version of ACG at your fingertips. The printed 500-page reference edition teams up with this online guide to give you hundreds of charter operators and brokers worldwide. The index conveniently sorts: charter operators by location, charter operators by name, U.S. brokerage services, and international brokerage services. There's even an interactive aircraft search to find your exact aircraft.

Although the printed edition tells the complete story about each operator, the online "zine" makes retrieval of basic info easy. Also on tap: charter industry info, discussions, and news. Industry reference topics give you a peek into: planning a charter trip, pricing air charter, the air charter broker, industry associations, in-flight catering, chartering an air ambulance, and glossary of air charter terms.

The site is expertly organized, fast, and helpful. Chartering? Dial this one up!

Fee or Free: Free.

PILOT RESOURCES

Introduction to GPS Applications
http://galaxy.einet.net/editors/john-beadles/introgps.htm

RATING
++

BRIEFING:
Position your browser here to find worlds of info on global positioning systems.

Aimed at new users of global positioning systems (GPS), this site voluntarily offers up great insight served in a plain brown wrapper. But, faster than you can say "spaced-based radio positioning systems," you'll be overwhelmed by GPS topics, issues, definitions, Q&As, products, and more. Congratulations to the author for such a monumental informational effort. If you're expecting graphics or English perfection, your cyber-hopes will be dashed. You will, however, find lots of pages relating to global positioning systems.

Just sifting through the loosely organized list, you'll discover there's more to GPS than push-button navigation. Highlighted subjects include: How GPSs Work, GPS Acronyms, Definitions of GPS Terms, Types of GPS Receivers, Policy Issues, Industry Applications (agriculture, surveying, photogrammetry, etc.), The NAVSTAR Signal Specification, GPS Satellites, and more. All online documents are color-coded with symbols representing the status of the document—from substantially complete to a proposed, but nonexistent page.

You'll learn about the author and his qualifications, find links to frequently asked GPS questions, unearth an archive site for GPSs, and be presented with sponsorship opportunities. Get ready to take notes, this online GPS class is in session.

Fee or Free: Free.

PILOT RESOURCES

Aero-Tourism

http://yi.com/home/RogerEtienne/aerotourism

RATING

BRIEFING:
An organized treasure chest of practical info for globally-minded flyers.

Thinking about taking the Cessna 150 to Switzerland? Okay, bad example. But, for those who like to break out of the familiar and spread their wings globally, Aero-Tourism is just the ticket. This site, simply designed for aerodynamic functionality, boldly avoids Uncle Jim's cross-country pictorials and limits the content to useful info.

The subjects are focused solely on global private pilotage. Topics include: foreign licensing procedures, places where one can rent, things to do, places to visit, and stories of flight abroad (caution: may need some multilingual skills here). The country-related info (as of review time) includes: Alaska, Australia, Canada, Columbia, France, Kenya, New Zealand, Portugal, South Africa, Switzerland, and Zimbabwe. Generally, the country resources are independent links to country-maintained sites. In almost every case, the content is useful and thorough. Specifically, you'll run into tidbits on overflight authorizations, ATC costs, civilian permits, prior notifications, The Aeronautical Information Publication (a global version of the AIM), airports of entry, and foreign exchange rates.

For fascinating facts and international flying insight, this is your first fix.

Fee or Free: Free.

PILOT RESOURCES

F.E. Potts' Guide to Bush Flying
Concepts and Techniques for the Pro
http://www.fepco.com/Bush_Flying.html

RATING
✝ ✝

BRIEFING:
Online book represents its printed predecessor with an unequaled reference for the bush pilot.

For those pilots who slip the surly bonds of crowded airspace, this unique online book should be a required flight manual. Wonderfully organized into a slick index and clickable chapters, the guide's content is crucial for bush pilots and fascinating for everyone else. This text-based resource is well-written (no fluff—just the useful stuff), with easy-to-follow tips and techniques. Generally, the topics and scope assume the reader carries at least a commercial ticket. But, everyone will enjoy a look at things like: partial stall landings, STOL landings, pre-heating, ground handling, and many images. The table of contents breaks down into the following categories: General Information, Equipment and Environment, Flying Techniques, Images, Glossary, Illustration, and Author Photo.

The only thing that unsettled me during my visit was the author's disclaimer: "I am primarily a pilot, not a writer…" To the latter I must disagree.

Fee or Free: Free.

PILOT RESOURCES

High Mountain Flying in Ski Country U.S.A.

http://www.tc.faa.gov/ZDV/hmf.html

RATING

BRIEFING:
A text-based summary of essential mountain flying info with specific emphasis on Colorado's ski country.

Tooling around the Rockies and other high altitude airports means extra sharp mountain flying skills. While expertly briefing pilots on the intricacies of Colorado flying, this educational site gives all mountain-bound pilots a great heads-up!

Brought to you by the concerned folks at the Denver Air Route Traffic Control Center and the FAA, this site's a great resource for explaining high mountain distinctions. Do's and Don'ts of Mountain Flying, in particular, is a well-written laundry list of hints, guidelines and safety essentials. Print it and keep it in your flight bag—it's excellent. Winter Flying is an equally informative masterpiece reminding pilots of the hazards of winter weather and aircraft operation. And, appropriately, there's a valuable refresher on density altitude.

If you're planning a flight into Colorado's Ski Country, check out these clickable topics: Commonly Flown Colorado Mountain Passes, Ski Country Airways Structure, Ski Country Airports, Flight Watch, Colorado Pilots Association Mountain Flying Course, and more.

Bookmark this site for safety's sake—it's a keeper.

Fee or Free: Free.

200 Best Aviation Web Sites

PILOT RESOURCES

The Hundred Dollar Hamburger—
A Pilot's Guide to Fly-In Restaurants

http://198.64.248.32

RATING

BRIEFING:
Looking for fly-in eats? This burger's well done!

Yes. I've mentioned the Hundred Dollar Hamburger before, linked to other sites. But, something this good deserves its own check ride.

Love to fly? Love to eat? Point the GPS toward the Hundred Dollar Hamburger—a Pilot's Guide to Fly-In Restaurants. Conscientiously updated, you'll get first-hand reviews and information on just about every general aviation fly-in worldwide.

First, click on your desired state or country (sixteen as of review). Second, choose the selected city to get an honest, straight-from-the-pilot's-mouth review. And, third, view your selection, complete with reviews. Write-ups are awarded one to five burgers, with five being best.

Be responsible. E-mail a PIREP of your favorite or not-so-favorite fly-ins.

Fee or Free: Free.

AirPage—The Interactive Aircraft Handbook

http://stega.smoky.org/~dlevin/

RATING
++

BRIEFING:
Specs, data, and histories galore cover close to 655 aircraft.

Trying to quickly compare a Citation II's range with a Lear 35's? Curious about the crew requirements for a Cessna T-37 trainer? Simply defined, the AirPage is a data intensive, interactive handbook. Review full descriptions, brief history, and technical data for each of the over 655 included aircraft. Although aircraft images weren't included until recently, the AirPage has amassed a huge selection of over 200 already. And, the Herculean task continues as you read this.

Currently, clickable categories include: Last Five Updates, Interactive Aircraft Comparison, AirPage Search, Aircraft Alpha List, Aircraft Maker Alpha List, Glossary, Aircraft Weapons, Aircraft Engines, Bulletin Board, and more.

Fee or Free: Free.

Webflyer

http://www.insideflyer.com

RATING

BRIEFING:
A knowledgeable resource for those collecting, cashing in upon, and seeking insight into the fine print world of frequent flyer programs.

Quite simply, Webflyer soars triumphantly into the murky abyss of frequent flyer (FF) programs. Skeptical? First, a few stats. At review time, Webflyer offered up over 800 pages translating into about 2,000 screens of pertinent info. There's 3,600 hyperlinks and 70 chat rooms. Second, the guy controlling the content has some experience with this sort of thing. How much? He's accumulated over five million frequent flyer miles/points, and conducted over 3,000 interviews on the subject with folks ranging from Congress to *Good Morning America*. Third, and most importantly, seeing is believing. Webflyer's organization and lofty style take flight with quick return menus, expertly tailored graphics, and nice subject area descriptions.

With so much to cover, I'll only tap into some highlights: The Top Ten List of Good Mile Deals; IF: NOW!–InsideFlyer's online, bimonthly "zine" that takes you into the heart of frequent flyer news; the Boarding Area lists and links every FF program known to man; and the Resources Area steers you toward more well respected reading on the topic.

More FF fun begins by clicking into Flyerfacts, Inside:Extras, Reviews, Flyertalk, Mile Marker, Funnies, and more.

Fee or Free: Free.

FlightSafety International

http://www.flightsafety.com

RATING

BRIEFING:

Company info perfection from world respected, high-tech aviation trainers—it's what you would expect from these folks.

What better way to emphasize a professional approach to training than with a top-notch site. FlightSafety's expertise seems now to stretch into the cyber-arena. The company's objectives and content descriptions are written well, avoiding long-winded sales pitches. Content links to FlightSafety particulars give you up-front and current information. There's a pop-up site index for quick page navigation. And, best for beginners, a colorful link menu points the way. Everywhere you go, though, you're always an e-mail icon away for more information on any topic.

Content links include: Courses and Schedules, Crew Resource Management, Maintenance Resource Management, Simulations Systems Division, Learning Center Network, People@flightsafety, News@flightsafety, and Jobs@flightsafety. Also: flip through Courses and Schedules to be introduced to the corresponding course material by aircraft manufacturer.

There's a reason why corporations, airlines, the military, and government agencies rely on FlightSafety. Tap into this site and you'll get a feeling why.

Fee or Free: Free.

GTE DUATS

http://www.gtefsd.com/aviation/GTEaviation.html

RATING

BRIEFING:
With free online flight plan filing and weather briefing info, this site should be every pilot's first stop.

Still one of the best resources in civil aviation today is GTE Contel's DUATS—Direct User Access Terminal Service. This valuable site provides current FAA weather and flight plan filing services to all certified civil pilots. The service is available 24 hours a day, seven days a week at no charge to the user—fees to operate the basic GTE Contel DUAT service (weather briefing, flight plan filing, encode/decode) are paid by the FAA. With DUATS you'll access completely current weather and NOTAM data. Instantly select specific types of weather briefings: local briefings; low, intermediate, or high altitude briefings; and briefings with selected weather types. The DUATS computer also maintains direct access lines for flight planning filing. You can file, amend, or cancel flight plans.

A tad computer shy? Never used DUATS before? No problem. Help areas take you by the hand with: Connecting, How to Log On, Weather Briefings, Flight Plan Filing, Entering Data, and more. If you fly... If you're a pilot... If you file flight plans... If you require weather info... Look no further and make room for a bookmark.

Fee or Free: Free.

Bookmarkable Listings

Simcom Training Centers
http://www.simulator.com
Simulator training programs and information.

GoldenWare Travel Technologies
http://www.traveldesk.com
Free travel information including hotels and car rentals.

Shareware Aviation Products
http://www.look-up.com
Aviation software tools for flight planning and management.

Aerodynamics and Flight Simulator
http://www.web-span.com/afs/
Information and demonstration for aerodynamics and flight simulator software.

Exotic Aircraft Company
http://www.barnstormers.com
Tips, procedures, and information on restoring vintage aircraft.

Airwise Hubpage
http://www.airwise.com/Navigate/front_page.html
An independent guide to worldwide airports and aviation/airline news.

Interjet
http://www.interjet-osi.com
Find charter aircraft available worldwide, plus up-to-date deadhead availability.

Best AeroNet
http://www.bestaero.com
Business aviation fuel network with jet fuel uplifts at over 700 sites worldwide.

Paul Tarr's GPS WWW Resource List
http://www.inmet.com/~pwt/gps_gen.htm#intro
Huge list of global positioning system (GPS) links and related references.

The Homebuilt Homepage
http://www.azstarnet.com/~cmddata/homebuilt
Central reference to homebuilt/experimental-class aircraft.

Aviation Museums, Education, & Flight Schools

National Warplane Museum

http://www.warplane.org

RATING
✈ ✈ ✈

BRIEFING:
The V-77 Stinson? Alive and well thank you, via this online museum.

Physically located at Genesco, New York, and on virtual tour via cyberspace, the National Warplane Museum site is captivating. Graphically motivated, behind-the-scenes buffs have obviously zeroed in on exceptional design. Better still, this mix of visual treats doesn't make a mockery of your modem. Loading is fairly efficient, multiple types of menus are always at hand, and the "cockpit" aircraft searching feature is fun.

Sure there's the usual on tap, like: Museum's Mission, Sorties (where the Museum's planes have been and where they are moving around to), Enlistment Form for Members and Volunteers, Tours Available, Pilot Interviews & Stories, Links to other sites, and more. But where you're going to want to spend some time is the online plane collection. Just find your favorite in the pop-up window and enjoy. From an F-14 Tomcat to an L-3 Grasshopper, it's all here via mouse and modem.

Fee or Free: Free.

Aviation Communication

http://www.flightinfo.com

RATING

BRIEFING:
Summed up succinctly as: "serving the aviation community, as well as potential flyers who just want more information on aviation."

Brand new at the time of review, Aviation Communication refreshes the experienced and preps the beginners with an information-rich heads up. You'll fall smack into insightful tips and solid info from seasoned pros. I even copied a few noteworthy tricks in Rules of Thumb, reminisced at the good advice in Checkrides, and refreshed myself with distances and separation in Airspace.

Don't expect online flying manuals or serious studying. This great source simply highlights and summarizes important topics. For example, Rules of Thumb combines helpful suggestions regarding: descent, ground speed, wind computation, bank angle, true airspeed, horsepower, pressure altitude, temperature, climb, instruments, and airworthiness.

Although there's not enough room here for details, you'll need to trust me and spend time clicking into: Learn to Fly, Legal to Log?, Message Board, Aviation Medicals, Post Resumes, Airline Addresses, Classified Ads, Flight Schools, Instructing Hints, and more.

It's aviator info euphoria. Enough said?

Fee or Free: Free.

The Hangar

http://www.the-hangar.com

RATING
✈ ✈ ✈ ✈

BRIEFING:

Hang out here for images, history, and hangar talk.

Step into this spacious hangar housing a delightful blend of history, images, and pilot career banter. Mostly steering towards worldwide military aircraft, your nicely framed selection features: Images of the Week—display current week winners and links to favorite past selections; Aircraft in the Hangar—use a pull-down menu to select by alpha characters, type, or nation; Hangar Talk—join fellow enthusiasts in this lively bulletin board forum; What's New(s)—scan a list of additions and site modifications; Historical Calendar—just select a date with this cool, searchable history book of military aviation events; Comments/Suggestions; Odd Bird Quarterly—learn about unusual aircraft; Search!—quickly locate site contents; Reviews—read comments about the site; Web Links—search the list of quality links; and Help/About—find tips and tricks for site navigation.

For me, subtle details pave the way for this award-winning mention. Stuff that earned the points: great descriptions next to links, searchable menus, and well-thought-out hints to save you time.

Fee or Free: Free.

Be a Pilot

http://www.beapilot.com

RATING

BRIEFING:
Dreamt of learning to fly? Fantasized about a well-organized intro-to-flight site? Touch down here for some encouraging reality.

Hey, stop dreaming and start browsing Be a Pilot's online taxiway to flight. Along with words of encouragement, you'll find visual page wizardry made up of handsome graphics that are quick to load. Omnipresent top and left-margin menus give you the guidance of a slick GPS. And plenty of efficient page layouts give you the "white space" to stay focused.

At the site's heart is an easy-to-use flight school search tool sortable by state. The surprisingly complete flight school list gives you many schooling options in your area, including brief descriptions and contact information. When you've narrowed your options and chosen a school, be sure to take advantage of the $35 introductory coupon. Just fill out the registration form (fairly lengthy) and print out the introductory flight coupon.

Other topics to call upon include: Welcome to Flying (nice, encouraging intro), What's New, Aviation Links (companies involved with and endorsing Be a Pilot), and the helpful Flying Library. The Library gives you a quick but accurate peek into safety, steps to getting your license, costs, and more.

Fee or Free: Free, but make sure to fill out the survey and receive a $35 intro to flight coupon.

Applied Aerodynamics: A Digital Textbook

http://aero.stanford.edu/OnLineAero

RATING
+++

BRIEFING:
A scholarly online lecture that gives aerodynamic buffs a lift.

Kudos to Professor Kroo and the rest at Stanford University. This digital textbook arouses the awe of aerodynamics for aviators worldwide. Intended to supplement a more conventional aerodynamics textbook (the printed variety), the online version of Applied Aerodynamics peeks into some winged wonders via the Web.

Similar to a textbook, this digital resource provides thoughtful site organization with right-margin topics, detailed table of contents, instructions, and index. Although graphic design gurus weren't called upon to send viewers into a GIF frenzy, the pages are simple and clean. Taking the place of visual perfection, interactive attributes within the text takes the form of analysis routines that were directly built into the notes. (See, for example, the streamline calculations, airfoils, wing analysis, and canards.) You'll also stumble across cool charts and various depictions throughout.

Intrigued as to topics covered? Here's a taste: fluid fundamentals, airfoils, 3D potential flow, compressibility in 3D, wing design, and configuration aerodynamics. Sounds like pocket protector stuff? Trust me and jump into the slipstream—it's drag-free.

Fee or Free: Free.

Aviation Ground School

http://www.geocities.com/CapeCanaveral/3819

RATING
++

BRIEFING:
Here's my favorite kind of ground school—simple, well-versed, and available 24 hours a day.

If you're like me, the actual wheels-up instruction was always more fun that memorizing FARs in ground school. It probably stands to reason that we could all use a tiny refresher now and then. Though not volumes of rocket science, Aviation Ground School gives your clicking fingertips a pointer to solid info vital to private pilots and wanna-bes.

With a simple, easy-to-understand manner, our educated author provides online instruction for a variety of topics. Particularly well written with charts and diagrams are: Components of the Fuel System; Aerodynamic Principles; Effects of Air Density on Performance; Airspace Definition, Airport Types, and Flight Rules; Aeronautical Chart Symbols and Their Meanings; Navigational Methods; Principles of Weather; Terminology and Procedures of Radio Communications; and more.

After you shuffle a few site typos under the rug, the overall site organization is pleasantly simple and quick to load. Mostly text-based, the information employs appropriate diagrams and avoids gratuitous pics. After clicking into a subject you may either move on to the next topic, or return to the index.

You will find a few other links here, but the online instruction is your real reason to stay.

Fee or Free: Free.

SimuFlite Training International

http://www.simuflite.com/main.htm

RATING

BRIEFING:
Advanced center for professional simulator training checks in with their own advanced promo site.

I'll readily admit the reason I initially visited SimuFlite's site didn't include simulator training. I clicked in for the link to live conversations between air traffic controllers and pilots in the Dallas-Fort Worth (DFW) area. Of course, now there are many ways to get to the DFW live link, but I must say taxiing through SimuFlite's site opened my eyes to their quality offerings.

For those interested in simulator-based training, SimuFlite straps you into their cyber-stopover for a well-designed intro. Right margin frame links team up with a non-frames version to bring any browser a closer look into SimuFlite's professional resources.

Clicking into any of the following exemplifies the site's layout and organizational skills: Logbook, About SimuFlite, What's New, Press Releases, Aircraft/Simulators, Training Schedule, Aviation Links, SimuFlite People, Employment, and Information Request.

Even if you're not SimuFlite-bound, do visit the weekly Ground Chatter collection of thoughts, sayings, and trivia. Depending upon your capacity for nonessential entertainment, it might just be worth a bookmark.

Fee or Free: Free.

Air Times

http://www.airtimes.com

RATING
++

BRIEFING:
A thorough look into the airline world—past and present.

Although its subtitle alludes to history, this airline-specific site pleasantly surprises with breaking news as well as historical highlights. You won't be "wowed" with design by anyone's standards, but the fantastic depth of info shines anyway. The archaic laundry list of hyperlinks almost seems apropos. A few "new" tags and links back to the top of the list comprise the only attempt at organization.

You may want to stick around for content, however. Updated daily, you'll tap into current industry news as well as archived press releases. A smattering of online annual reports also peeks into the more recent trends and ramblings among the big jet pros. Then, reminisce with articles on airline history. Read about Transamerica Airlines Corporation (text and graphics from 1932). Or, delve into a detailed photo history of California's largest intrastate carrier—Pacific Southwest Airlines.

Get comfy here with a favorite beverage—the captain has turned off the "fasten seat belts" sign.

Fee or Free: Free.

The Aviation History On-Line Museum

http://www.aviation-history.com

RATING
✦ ✦ ✦

BRIEFING:

Tour this online museum for a cyber-stopover into the archives of aviation history.

Aviation enthusiasts young and old will revel in the online convenience of this interactive museum tour. No dusty textbooks here. Just pixel-quality pics and logs of insightful description. Those with a fancy for flying will quickly become enamored with a nice collection of historic aircraft.

Neatly indexed by manufacturer and country, a quick click launches you into a concise aircraft summary—with an option for the full text version. Although the punctuation-challenged descriptions run amok, the facts and stories are still fascinating. The handpicked index of aircraft includes more than a few of my favorites: P-26 Peashooter, Mosquito, P6 Sea Master, Spitfire, P-47 Thunderbolt, B-24 Liberator, P-40 Warhawk, and The Concorde. Yes, even the Corcorde. Hey, it's still a history-making phenomenon.

Once you've saturated your curious mind with aero wonders, there is still more to study. Mouse over to the left-margin windows for: Aircraft Engines, The Early Years, Construction Technology, Theory of Flight, *Aviation Magazine*, Fantasy of Flight, and more.

Fee or Free: Free.

See How It Flies

http://www.monmouth.com/~jsd/fly/how

RATING

BRIEFING:
Not your standard book on flying airplanes, this new spin on piloting gives even veterans a reason to jot a few notes.

Set aside your AIM, the FARs, and your operations manual for the moment. Buckle into the right seat for a little online instruction from FAA safety counselor, John Denker. Then, get ready to learn something—even if you think you know it all. See How It Flies might surprise you with thought-provoking commentary on the perceptions, procedures, and principles of flight.

Easily arranged in table of contents format, See How It Flies packs unlimited tips and worthwhile piloting techniques into a refreshingly candid resource. Clicking into the chapter file folders unearths no-nonsense tricks and useful procedures to help you fly better. Learn how to use your eyes, ears, fingertips, and seat of your pants to gather info. Find out about using your hands and feet to make the airplane do what you want. And, read about how to organize your thinking.

Well-written, practical prose skims the surfaces of: vertical damping, roll damping, and stalls; angle of attack, trim, and spiral dives; slips, skids, and snap rolls; take off; landing; cross-country flying; and more. You'll find chapter contents are easy to understand with well-defined concepts and corresponding diagrams. Thankfully absent, however, are meaningless plane pics and inappropriate graphics. So, enjoy and learn in frustration-free peace.

Fee or Free: Free.

Air Forces of the Americas Almanac
http://www.cdsar.af.mil/almanac/english/enghome.html

RATING
++

BRIEFING:
Look up the Americas' air forces on a virtual silver platter via this collection of unclassified data.

Just as most almanacs lean toward a dry, facts-only style, the Air Forces of the Americas Almanac is no exception. Its contents zero in upon the current air force stats fluff-free. The only fancy site element may be found in depth of content. Scanning the well-researched breakdowns and tables causes you to forget about the lack of visual niceties.

The main offerings presented in your choice of English, Spanish, or Portuguese feature: an air forces index by country; U.S. Southern Command info; a descriptive peek into the System of Cooperation among the American Air Forces (SICOFAA); and a total population versus air force members table. Most fascinating, in my opinion, is the country itemization of air forces. The index allows for 27 clickable countries, including places like Honduras, Mexico, Belize, Chile, Jamaica, and the U.S. Once a country is chosen, the detailed air forces dissection begins. Read about stats for: members breakdown, headquarters, commander, air bases, and more. Equipment statistics delve into type and quantities of aircraft used as bombers, transport, fighters, tankers, training/reconnaissance, and other missions.

Fee or Free: Free.

Helicopter History Site

http://www.helis.com

RATING
+ +

BRIEFING:
Though not a scholarly resource, this rotor wonder offers some fun pictorial history.

Whirling in a virtual fog of shaky English translation and painful layout, the Helicopter History Site emerges unscathed as an award-winner nonetheless. Carefully navigating through the random ad banners and slow-to-load pics, you'll stumble across a surprising collection of helicopter history. Your first tip is to gather ample amounts of patience and understanding.

Setting aside for the moment any hope of design or typo-free description, navigating within the site is actually easy. Thanks to a main menu content index by timeline or company and handy continuation links, maneuverability resembles that of a Bell 430. Stepping through the decades of manufacturers, models, and inventors, the historical tour holds your hand through many decades. Although descriptions are brief, the pictorial reviews are good, showing a nice "helio-progression." From Leonardo Da Vinci's Helical Air Screw to the Bell/Boeing 901 Osprey (V-22), if you're a helio-buff raise your level of history knowledge here.

Fee or Free: Free.

TheHistoryNet Archives—
Aviation and Technology

http://www.thehistorynet.com/THNarchives/AviationTechnology

RATING
✈✈✈✈

BRIEFING:
Gather around the monitor for some good, oldfashioned aviation tales.

Stemming from the first-rate excellence of TheHistoryNet, the Aviation and Technology archives promise a well-documented review into the history of aviation.

Unlike the drab presentation often associated with historical reference, TheHistoryNet's insights come alive with color and style. Obviously well-researched, the writing is flowing and interesting—a key ingredient in successful informational ventures. Photos and illustrations are many, but none so large that modem time is compromised. Each article even offers a one-page summary with a link to the full text.

For a history site covering more than just aviation, TheHistoryNet's depth of aero articles is impressive. I saturated my brain with features like: "Airmail's First Day;" "The Guggenheims, Aviation Visionaries;" "Kalamazoo 'Air Zoo';" "Luftwaffe Ace Gunther Rall Remembers;" and "Stealth Secrets of the F-117 Nighthawk."

When you've finally reached the end of the archived Aviation and Technology list (50+ entries), a convenient left-margin index invites you to explore other historical topics. Although these have nothing to do with aviation, the invitation still stands.

Fee or Free: Free.

US Air Force Museum

http://www.wpafb.af.mil/museum

RATING

BRIEFING:
Stop by for Air Force specs aplenty—you'll be bombarded with bombers and inundated with insight.

It's a good thing that superb site navigation, in the form of a left-margin menu, takes you by the hand. Without its directional beacon you'd be lost in a virtual sea of bombers, trainers, and fighters. The sheer volume of images and history falls into the unbelievable category. Plan for extensive viewing if you're an Air Force admirer.

While I'm certainly not trying to diminish the importance of the actual museum in Dayton, Ohio, this promo site is quite a spectacle in itself. In my opinion, one of the best starting points for a mega-site like this, is to take some virtual tours. My favorites are the Century Series Fighters (the F-100 through F-107 jet fighters), The Presidential Aircraft Collection, and WWII Air Base Operations. Then, start browsing through the multitudes of aircraft and special galleries. Modern Flight, Early Years, Air Power, Space Flight, and others present fantastic capsules of memories. From History to Engines & Weapons, there's more than a couple days worth of viewing here alone.

Fee or Free: Free.

National Air & Space Museum (NASM)

http://www.nasm.si.edu

RATING

BRIEFING:

The Wright Brothers would be proud to take this cyber-museum tour. Although the online museum is fascinating in pixel version, don't let it quell your thirst for seeing the real thing.

It's simple and informative—the way a world-class museum ought to be. The Smithsonian Institute's online National Air & Space Museum gives you a virtual look at aviation and space history. Click through museum maps and exhibits, educational programs, NASM news and events, NASM resources, or just general information about the museum itself. A convenient, clickable museum map points you in the direction of your favorite exhibits. From Milestones of Flight to Rocketry & Space Flight, you'll scan through online gallery greatness.

Have a specific question or winged fancy? Just jump into the powerful search engine. Search the Smithsonian Web by typing your phrase(s) or keyword(s). It's history at your fingertips.

Fee or Free: Free.

Aviation Web

http://www.aviationweb.com

RATING
+ + +

BRIEFING:
When your flights of fantasy have turned to something more than dreaming, check into this inexhaustible U.S. flight school directory.

Hey, these guys concentrate on one thing: flight schools. There's not a lot of extra stuff that throws you off course. When you're ready to move into a twin, wrestle around a Boeing 777 commercially, or just flutter around in a trusty 152, climb into the left seat with Aviation Web. This searchable database easily finds your area's best schools via three criteria: city, state, and/or zip.

Certified flight instructors and pilot schools are invited to create an online company or personal profile FREE through an update form. This is a great resource designed for an international base of pilots (and pilot wanna-bes).

Fee or Free: Free.

San Diego Aerospace Museum

http://www.AerospaceMuseum.org

RATING

BRIEFING:

Dazzling aviation history through a futuristic Web site tour.

Wow! High resolution pics, well-written history, and perfect organization catapult the online edition of San Diego's Aerospace Museum into a must-see site. Aviation enthusiasts and historians are in for an exhibit tour encompassing the Dawn of Powered Flight through the Space Age.

True, online is a nautical mile from the real thing, but begin your journey here. There's hordes of fascinating info. Delve into the Montgolfier Brothers' Hot Air Balloon of 1783 (the first manned vehicle in recorded history to break the bonds of gravity), or read about your favorites: Lindbergh, Earhart, Gagarin, Armstrong, and more. Museum hours, fees, phone number, location, collection listing, and special event services are just a click away also. If you're just revisiting, quickly find new additions in: What's New at the Museum, Education Programs, and Library/Archives.

Recommendation? See the online version, then be dazzled by the real-life stuff.

Fee or Free: Free to view, fee for in-person museum tour.

Neil Krey's Flight Deck

http://web2.airmail.net/neilkrey

RATING

BRIEFING:
Welcome aboard the Flight Deck for an educated view on training and learning.

Whether you're relying upon the instructor in the right seat for guidance or an airline pilot to get you home, experience factors into the background of most longtime pros. Captain Neil Krey of Neil Krey's Flight Deck site has quite a background. This man, who is in the business of training and education programs, is an accomplished commercial pilot with an eye toward many aviation research areas: crew resource management, training & learning, the future, and more. In the Flight Deck you'll find links to important topics, such as: Web-based Training, Aviation Safety Reporting System, and Scenario-based Planning.

There's an educated look into aviation's future, as well as many fascinating studies and papers published by Captain Krey. Make yourself comfy in the jumpseat—you're going to learn something here!

Fee or Free: Free.

The Aviation History of Wichita, Kansas
The Air Capital

http://www2.southwind.net/~wknapp/air_cap

RATING

BRIEFING:
Enjoy masterful navigation through the early years of The Air Capital.

You're thinking that aviation history means textbooks, microfiche, and Great Granddad's yellowing photos. Well, don't reach for your library card just yet. Explore the breakthrough aviation beginnings in Wichita, Kansas (The Air Capital) right here at the Aviation History of Wichita's home page.

Deftly organizing the early aircraft years (1911-1929), this site offers wonderfully written accounts and informative historical facts. The mostly text-based history does include some fun photos and graphical representations. A series of chapters chronicles Wichita's flying fancy: The Beginning of a Love Affair; Flying by the Seat of Your Pants; Wichita's Father of Aviation; Building Planes and All That Jazz; Travel Air—A Major Success; Trouble in Paradise; Cessna Sets Out Alone; Stearman Returns; Wichita—The Air Capital; and The Rest as They Say is History.

It's the perfect place to relive the lives of aviation greats like Clyde Cessna, Eugene Ely, Earl Rowland, Walter Beech, and more. Your thirst for aviation knowledge ends here—it's an oasis.

Fee or Free: Free.

University of Nebraska at Omaha (UNO)
Aviation Institute
http://cio.unomaha.edu/~unoai/aviation.html

RATING

BRIEFING:
Recruiting tool for UNO's Aviation Institute with an eye toward useful info for non-aviation-education seekers.

This higher learning site soars with knowledge not only relating to aviation, but effective self promotion as well. Gliding effortlessly through the Aviation Institute's pages, you'll realize that all of its wisdom, visually appealing style, and useful information coagulates into a nicely packaged recruiting presentation. If you're in the market for lofty education—great—begin your search here at University of Nebraska at Omaha's Aviation Institute. Take a sneak peek into course lecture notes, programs and projects offered, careers in aviation, key reasons to enroll, Institute background info, etc. Explore undergrad and graduate programs, flight training, or an aviation minor.

Non-education-bound browsers will find the info here useful as well. A Flying and Weather page brings you information on flying techniques for pilots, flight planning forms, weight & balance calculations, fuel and E6B info, FAR/AIM resources, and weather links. Oh, and be sure to check out the Aviation Institute-sponsored Journal of Air Transportation World Wide. It's packed with articles and online dialog from many experts.

Fee or Free: Free.

A Virtual Museum Describing the Invention of the Airplane

http://hawaii.cogsci.uiuc.edu/invent/airplanes.html

RATING
✝✝

BRIEFING:
A cyber-museum for aviation invention enthusiasts.

You don't have to keep your voice down or worry about knocking anything over in this museum—it's virtual. Its beneficial contents, however, are tangible and inviting. Fun graphics and sporadic quotes make learning via computer an online adventure. Step into the museum for: The Tale of the Airplane (a Puritan Fairy Tale); The Design and Test Strategy of Invention, The Dark Unhappy Ending; The Photo Gallery; and The Bibliography (a list of relevant readings for those who wish to learn more). Also on cyber-tap are special features like current public service announcements and various other historical stuff.

No, it doesn't substitute for a live walk around of The Spruce Goose or The Wright Glider. But, the enjoyment and learning potential is as real as can be in this virtual wonder.

Fee or Free: Free.

FirstFlight

http://www.firstflight.com/index.html

RATING

BRIEFING:
No plane. No stalls. No talking to the tower. No expense. It's the cyber-way to pilot a Cessna 152 for the first time.

Although a crafty approach to enticing new recruits, FirstFlight wildly succeeds in captivating potential flyers with some fun cyber-152 instruction. Excellent content and organization make FirstFlight easy to use for Net novices as well as aviation novices. Unlike a flight simulator, this site steps you through the private pilot certification process via a series of "flights." Although not intended as a substitute for actual instruction, the evolving list of "flights" (new ones are added monthly) are a great preview to the real thing.

Scan through the current private pilot requirements. Read the private pilot syllabus. Examine checklists. Take an online preflight. And strap into the left seat. Following the script in each "flight," you'll become familiar with your cyber-Cessna: taxi, communicate via radio, take off, and land! You'll encounter embedded checklists throughout each "flight"—they're excellent references. Much of the information is basic and well suited to a flying introduction.

Congratulations to this innovative, educational, and interesting site. (How many sites can claim all those wonderful adjectives?) This guy's good...where do I sign?

Fee or Free: Free.

Embry-Riddle Aeronautical University

http://www.db.erau.edu

RATING

BRIEFING:
A perfect demonstration of how smart folks (well, they are aviation educators) can grab the technological reigns and capitalize upon appropriate Web applications.

From the minute you type its address, the Embry-Riddle page exceeds your expectations—even from the world's largest aeronautical university. This online equivalent of a university brochure reeks of outstanding aesthetics while satisfying content-hungry surfers. Page navigation is simple with well-designed, omnipresent menus and clickable illustrations. Also available is a thoughtful, low-graphics version for less speedy computers and/or modem connections.

Main topics are clearly labeled as: Visitors, Students, Information, Research, Faculty/Staff, and Administration. Its content, quite simply, overflows the cup of perfection. From university info to research links to aviation links, you could easily spend days here. Education seekers will find admissions stuff, financial aid info, and a cool clickable campus map tour. Students (and anyone else) can tap into Career Info, the Avion Online (University-sponsored "zine"), and Associations. There's general campus news, library info, and more research areas. You'll even have at your fingertips: faculty/administration info (including phone numbers), colleague information, and more.

In person or online, this university offers quite an education.

Fee or Free: Free.

Amelia Earhart

http://www.ionet.net/~jellenc/ae_intro.html

RATING
✈✈✈

BRIEFING:
A brilliantly orchestrated Web tribute to America's most famous aviatrix, Amelia Mary Earhart.

Delving into the wonderfully ambitious world of Amelia Earhart, this cyber-tribute justifies every one of the awards it has garnished. Expertly written, illustrated, and presented, this online tour of Ameila's courage takes the browser through: The Early Years, The Celebrity, and The Last Flight. The fascinating text is easy to read and insightful. Clickable photos are scattered throughout, as well as clickable icons that take you into each chapter.

Included with Amelia's story are a few extras. Scan through unconfirmed themes as to her mysterious disappearance. Or, browse info regarding the Earhart Project—an investigation launched in 1988 by The International Group for Historic Aircraft Recovery (TIGHAR) to conclusively solve the mystery of Amelia's disappearance.

For those wanting to continue the education, tap into the site's film links to Flight for Freedom, Amelia Earhart, and Amelia Earhart: The Final Flight. Other related links include: Discovery Gallery, Tall Cool Woman, The Sky's the Limit, Famous Women in Aviation, People's Sound Page, Mystery of Amelia Earhart, Howland Island, and The Ninety-Nines.

Fee or Free: Free.

TIGHAR

http://www.tighar.org

RATING

BRIEFING:

Fascinating historical research, masterfully displayed by the world's leading aviation archaeological foundation.

Pulled from the deepest, darkest caverns of the Web's cyber-cellars, TIGHAR (The International Group for Historic Aircraft Recovery) has, itself, been discovered. Not familiar with this diamond-in-the-rough? The nonprofit organization happens to be the world's leading aviation archaeological foundation. Their goals of finding, saving, and preserving rare and historic aircraft are artfully displayed here—online.

Fascinating history presented through well-written and descriptive pics are at your disposal. Read through a thought-provoking investigation into the disappearance of Amelia Earhart in The Earhart Project. Learn about the disappearance of Nungesser and Coli aboard l'oiseau Blanc in Project Midnight Ghost, and probe into rumors that WWII German aircraft still survive in underground bunkers in Operation Sepulchre.

Also available: historic preservation articles, other resources, and a look into the TIGHAR Tracks Journal.

Fee or Free: Free to view. Regular, student, and corporate TIGHAR memberships available.

Dryden Research Aircraft

http://www.dfrf.nasa.gov/PhotoServer/photoServer.html

RATING

BRIEFING:
Researching research aircraft? Or, just need a couple of fun copyright-free photos? Here's your site.

A dizzying array of digitized delights are housed here at the Dryden Research Aircraft Archive home page (physically located at the NASA Dryden Flight Research Center at Edwards, California). With over 600 images, the archive offers a huge selection of research aviation photos dating back to 1940. No copyright protection is asserted for any of the photos and multiple resolutions are available: 1280 x 1024 and 602 x 480; all new images are also available in 2000 x 1720 (8 ½ x 11 x 200 dots per inch). All are even 24-bit color JPEGs.

Site info resources include: What's New, Top Ten Photos Downloaded by Month, Dryden Movie Gallery Page, Dryden Fact Sheets Page, Dryden Flight Research Projects Page, and The Dryden WWW Home Page.

From the B-47 Stratojet to the F-14 Tomcat—it's all a keystroke away.

Fee or Free: Free.

Learning to Soar

http://acro.harvard.edu/SSA/articles/learn_soar.html

RATING
✈ ✈ ✈

BRIEFING:
A nondescript gem of useful, GIF-less information about learning to pilot a glider.

Hidden deep in the darkest reaches of the Web, way past the glitzy bandwidth-hogging pages, you'll find Learning to Soar. Shunning the hindrance of noisy engines, this soaring site sets you free to discover everything you've always wanted to know about becoming a private glider pilot.

The text-only information is accurate and thorough. The author spends a tremendous amount of time stepping you through the entire private glider pilot process. Topics include: glider ports, gliders and instructors, minimum training requirements, medical requirements, student pilot certificates, the written exam, training schedules, the flight test, total costs, glider vs. airplane training, private pilot privileges, and a few important soaring contacts.

Sidestep the search engines (and those graphic-intensive sites) to find glider knowledge aplenty here. It's an uplifting experience.

Fee or Free: Free.

Bookmarkable Listings

Port Columbus Historical Society
http://www.asacomp.com/~tkeener/pchs.html
History on airliners scheduled into Port Columbus since 1929.

106th Rescue Wing
http://www.infoshop.com/106rescue
Info, history, and recruitment details for the 106th Rescue Wing of the New York Air National Guard.

390th Bombardment Group
http://www.airfax.com/390th
Memorial museum preserving the proud heritage of the 390th Bombardment Group.

The College Aviation Resource Page
http://junix.ju.edu/HomePages/Aviation/avlinks.html
An educational resource guide for college aviation students, faculty, and flight enthusiasts.

Virtual Tour of the Boeing 727 Cockpit
http://www.msd.org/727.htm
The Museum of Scientific Discovery's virtual tour of the Boeing 727.

GG-Pilot
http://www.gg-pilot.com
Directory of America's top flight schools.

Aero Data Files
http://www.tcsn/adf
Free, online reference material for aviation historians, researchers, writers, and scholars.

UND Aerospace
http://www.aero.und.edu
News, resources, contacts, and details regarding University of North Dakota Aerospace.

History of the Brazilian Air Force
http://www.mat.ufrgs.br/~rudnei/FAB/english.html
Info, stats, airbases, units, and more relating to the Brazilian Air Force.

College of Aeronautics
http://www.aero.edu
Cyber-hangar for the College of Aeronautics at LaGuardia Airport in New York.

Aeroflight
http://www.netlink.co.uk/users/aeroflt/index.html
Find detailed profiles of lesser-known aircraft types, info on NATO and non-aligned European air forces, and more.

Luc's Photo Hangar
http://www.bayarea.net/~hanger
Pictorial view of World War II aviation history in the form of aircraft nose art.

8th Air Force in World War II
http://www.collectorsnet.com/milhist/index.html
Online informational tribute honoring the members of the 8th Air Force.

Rhinebeck Aerodrome Museum
http://www.oldrhinebeck.org
Sneak peek into this living museum of antique aviation.

Air Cruise America
http://www.aircruise.com
Nostalgic tour of the DC-3, complete with technical specs, type ratings, and more.

The International Women's Air & Space Museum
http://www.infinet.com/~iwasm
Online museum preserving the achievements of women in aviation.

CFI Central
http://www.cficentral.com
Nationwide search tool for certified flight instructors.

The Spruce Goose
http://www.navicom.com/~gandalf/spruce1a.htm
Unofficial page delves into the history of the Spruce Goose with photos, historical details, perspectives, and more.

American Airpower Heritage Museum
http://www.avdigest.com/aahm/aahm.html
Official pictorial tour and information relating to the American Airpower Heritage Museum of the Confederate Air Force.

The Air Base
http://www.airforce.com
U.S. Air Force recruiting site and information center.

Aviation Online Magazines & News

CyberAir Airpark

http://www.cyberair.com

RATING

BRIEFING:
Touch down at The Park for tantalizing tidbits, prizes, entertainment, and handy safety stuff.

Once you land at CyberAir Airpark, you've got a myriad of stops right off the taxiway. Click on an Airpark map or select from the site index for: Control Tower and FAA Info, The Museum, The RealAudio Area (have your audio player geared up), The Press and Media Center, The Office Center, The Fixed Base Operations Building, Links, and more.

Although the above locations certainly warrant a visit, you may want to skip directly to Hot Stuff selections, including: Last Week's Aviation History, Free BFR at CyberAir Airpark, CyberAir Beacons—updated weekly, Aviation Safety Pages, and Aviation Trivia (monthly winners receive better-than-expected prizes).

Page navigation skims the surface of simple, with clickable map or index options. Nifty additions include a scrolling info banner and a link to Chicago Approach Control—live!

Fee or Free: Free.

The Homebuilder's Den

http://www.mosci.com/buildersden/main.html

RATING
✈ ✈ ✈

BRIEFING:
Stop by The Den to satiate your homebuilt curiosities.

Not necessarily on the Kitfox Vixen level of visual perfection, The Homebuilder's Den struggles with funky clip art but soars with content. Hopefully your visual tendencies will yield momentarily to allow your more logical side a taste of homebuilding insight.

All graphically oriented omission aside, The Den does exactly what information-based forums are supposed to do—inform. Frankly, I am thrilled with the lack of space-hogging plane GIFs. Navigation is simple—just scroll to the topic you want (how novel). And, the descriptions before you jump to a topic surpass any 250K illustration that's supposed to dazzle me.

At the time of review, my favorites (yes, there are many) include: My Weekend Adventure, The Kitplane Decision, Brass Monkeys, Fabric Application, Diary of an Aspiring Pilot, Panel Matters, Builder's Stories, Homebuilder's Laws (humorous), and Tool Time.

Fee or Free: Free.

AeroWorldNet

http://www.aeroworldnet.com

RATING
┼ ┼ ┼ ┼

BRIEFING:
International aerospace news that's updated weekly and written well—the perfect ingredients for a bookmark.

The global aerospace perspective—you'll find it online at AeroWorldNet. Current and updated weekly, this international aviation news source gives you efficiency and excellent insight. Topics covered: Feature Stories, Weekly Headlines, Industry News, Briefs in Aerospace, Behind the Scenes, and more. Great, short summaries capture the essence of most articles before you click into the full story.

Mostly text-based with some banner ads and logos floating around, the pages are quick to load and easy to navigate. Left-margin links take you to additional subjects like: Aerospace Jobs, People and Places, Industry Literature, Industry Products, Aerospace Events, Industry Message Board, Aerospace Companies, Industry Associations, and Membership (sponsorship) in AeroWorldNet.

With not many places to turn for solid, aerospace news online, you will be rescued when you land on AeroWorldNet's informational oasis. Lacking are a plethora of typos, giant pics, and old news. It's a world of positive difference compared to your other Web options.

Fee or Free: Free memberships—sponsorships are available.

Aviation Weekly

http://aviationweekly.com

RATING
✈ ✈ ✈ ✈

BRIEFING:
Web radio show just for us aviation types.

A radio show on aviation? Yes. But, better still, you don't need a radio and you can tune in live—worldwide! Self-dubbed as "the first and best," Aviation Weekly rides the radio waves of Web innovation and brings aviation enthusiasts this unique approach to industry news.

Traveling at 186,000 miles per second, Aviation Weekly's aeronautical "audiology" comes together into entertaining shows. Left button topics move you into your audio preferences of: Download RealAudio (free), Info About Page, Aviation Brunch (a fun, Sunday show featuring "aviation enthusiasts with attitudes"), About Aviation Weekly, and Listen to Aviation Weekly (live broadcast on Fridays 7 p.m. CT—0100 GMT).

Here's a taste of past shows: "B-2 Stealth and Aerospace," "Airline Outlook," "Best Aviation Products for the Holidays," "Aviation Jobs," and more. All are insightful and well-hosted.

Crank up your 14.4 kbs (or faster) modem and point your browser here to hear.

Fee or Free: Free.

General Aviation News & Flyer

http://www.ganflyer.com

RATING
✈✈✈

BRIEFING:
Online "zine" mirroring the printed one—no teasing here though, there's plenty of articles and features.

This online complement to its printed twin glides with effortless site navigation and hearty aviation news. Unlike other publications, which try to sucker-punch you into subscribing, this online sibling actually gives you the printed features, current articles, and commentary without teasing. Sure, they invite you to subscribe. But they're low-key about it.

Get to the heart of the site by clicking into Read the News. You'll tap into the current issue, as well as hordes of additional news brought to you by Landings (a giant info-rich site of its own merit, also in this book). Regular categories include: News, Business Briefs, Features & Opinions, Of Wings & Things, Alaska Flyer, Western Flyer, Sun 'n Fun, and Cessna & Mooney Mods/Maint. Want more general aviation? Click About GANews & Flyer, GANews & Flyer Archives, Search the Classifieds, Calendar of Events, RoBen Aviation Books, and a few links.

Fee or Free: Free.

Seaplane Pilots Association

http://www.seaplane.org

RATING

BRIEFING:
Have a seaplane fancy? Get your feet and your floats wet here.

Although fairly unusual in the Web world, Seaplane Pilots Association is mostly informational. Yes, that means no ads, banners, or membership driven hoopla. Just the water-flying facts.

And, if the lack of junky ads weren't reason enough to visit, Seaplane Pilots Association is efficiently organized too. You'll wonder why all aviation sites haven't copied its source code. Simple page links with nice descriptions get you into more water-flying fancy than one should be allowed in a Web visit.

Clickable contents include: Seaplanes—Overview, Flying Boats Overview, Flying Boats by Make & Model (many pics), Floatplane Overview, Floatplanes by Make & Model, Homebuilt & Ultralight Seaplanes, Floats & Mechanical Details (floats, land-to-floatplane conversion, and corrosion), and Seaplane Airlines/Air Taxi Directory (search by region or airline). A Further Information section points you in the direction of instruction, links, handbooks, and mailing lists. It's a one stop seaplane shop.

Fee or Free: Free.

Rotorcraft.com

http://www.rotorcraft.com

RATING
++

BRIEFING:
Here's enough info to keep your head and your gyrocopter spinning.

Grab the stick and go for a gyro spin with Rotorcraft.com. Probably not for everyone, especially dedicated fixed-wing folks, Rotorcraft.com does a great job of gearing you up with endless screens of gyrocopter info. The at-a-glance Reports section delves into current events and fly-in reports, as many more categories stand by in the margins. Scroll down to reveal your clickable options: Upcoming Events, Chapter News, Aviation & FAA Sites (only a few as of review time), International Organizations, PRA Connection, Manufacturers List, Instructors List, Classified, Gyro Store, New Products, In Development (homebuilders development page), and E-mail Lists & Expert Contacts. Among the gyro insight offered, you'll find answers to: How much do gyros cost?; Is a license required?; How much does training cost?; and How can I get a test ride?

Looking for more rotor fun? Rotorcraft.com happens to be the home server site for the Popular Rotorcraft Association, Air Command International, and Rotor Flight Dynamics, Inc.

Fee or Free: Free.

Business & Commercial Aviation (B/CA)

http://www.awgnet.com/bca

RATING

BRIEFING:
Business and commercial aviation "zine" that gives you more than just a hard sell into subscribing to its printed twin.

Fee or Free: Free, unless you subscribe to the printed version.

Unlike some online versions of printed magazines, Business & Commercial Aviation reaches beyond the GIF-driven solicitation for subscription. Sure, an online subscription desk is still ready for your order. But, you'll find substance here too—more than you might think.

Through an excellent labyrinth of well-positioned frames, news and opinions are everywhere. Well-written gems include monthly columns from legendary pros, current events, and in-depth feature articles. Your always-present topic menu moves you quickly into the mainly magazine topics of: B/CA Intelligence, B/CA Observer, B/CA Viewpoint, Safety, Top Stories, Columns, About B/CA, Subscription Desk, and Gallery.

Sure the articles below will change monthly, but to give you a taste read on. Lively discussions and professional analysis are at your clicking fingertips with articles like: "Reinstatement of Aviation Excise Taxes is Temporary;" "FAA Says Mid-Air Close Calls are Declining;" "All-Inclusive Warning System is Under Development;" and "1997 Salary Survey."

Also, a fun little sidebar features a gallery of images, videos, and sounds. The convenient mini search engine helps to browse for the latest gallery additions. I think you'll find the unending list of newsworthy GIFs to be pleasantly overwhelming.

200 Best Aviation Web Sites   **155**

FlightWeb

http://www.flightweb.com

RATING
++

BRIEFING:
An air medical must-see. Check in for current, thorough industry information.

Faster than you can say "E.R.," FlightWeb bursts onto your screen and serves up no-nonsense air medical resources. Homebuilders have their own sites. Aerobatic buffs have theirs. And, now the air medical pros have an online forum to gear up for air medical transport readiness.

This dedicated breed of specialists comes together with their own well-organized cyber-support staff. Not limited to just pilots, FlightWeb dips into content designed for worldwide air medical folks who are flight medics, nurses, medical doctors, communication specialists, dispatchers, and more.

Text-based menu links come complete with concise descriptions and offer a wide range of worthwhile reasons to add a bookmark. Industry professionals will want to scan through: Air Medical Web Pages—links to flight program Web pages worldwide; Chat Room; Flightmed Mailing List Archives—a searchable grouping of posted messages; FlightWeb Resources—employment listings, mentors, medical protocols, legal issues, and various FAQs; FlightWeb White Pages—e-mail directory of air medical pros; Associations; Vendors; and more.

Scrub up and join FlightWeb in this well-educated online operatory. The air medical doctor will see you now—24 hours a day.

Fee or Free: Free.

A/C Flyer Online

http://www.acflyer.com

RATING
✈ ✈ ✈

BRIEFING:
A great A/C Flyer "e-zine" that steers clear of "smoke-and-mirrors" information.

More than mere "e-zine" camouflage, A/C Flyer Online dutifully avoids the cheap subscription-only tease and serves up meaty aviation delights. This flashy electronic version of its popular sibling in print may surprise you with helpful formats and regularly updated news.

Well-organized search engines guide you to endless listings of aircraft, products/services, and dealers/brokers. Current market news is always conveniently at hand by clicking A/C Flightwatch. And, the informative Ownership Articles touch on many important topics. Aviation links, show news, image gallery, and a link to the Aviation Week Group round out your thorough browsing options.

Moving through the useful content is a snap. The subtle use of appropriate design elements balances nicely with useful navigation. Multicolored menu buttons help categorize your selections. And, main menu text links guide you like a finely tuned GPS. Sure, you'll need to dodge a few ad banners—but it's to be expected from advertising-based publications.

Fee or Free: Free.

Pilot's Web

http://pilotsweb.com

RATING
✈✈✈

BRIEFING:
Online-only magazine debuts with solid features and ongoing departments.

A fledgling pilot pub has broken through the cyber-ceiling, and I thought you might be interested to know. If you click in knowing beforehand that some surfaces might be a bit rough around the edges (design, content, and navigation), then your preflight's complete. Just keep in mind that getting around in the site means the use of "next page/previous page" buttons connected to a main menu interface.

Mirroring your favorite printed mags, Pilot's Web has the feature stories combined with ever present departments. Classifieds are proudly selling their wares in the back. And, archived issues are conveniently clickable too. The feature stories (during the reviewed issue) serve up pertinent flying info in the form of text-based articles, diagrams, and photos. Leaning more toward practical info, Pilot's Web will take you through feature topics like: Distance and Direction, Uphill and Downhill Take-offs, The Jetstream, Aviation Lawyers, and more.

Although you'll probably be tapping into future editions with changing features, you can always count on the ongoing departments of: Weather Sense, Flight Training, Calendar, Classified Ads, and Archived Issues.

Climb aboard with some extra surfing time on your hands. It may be relatively new, but it sure is airworthy.

Fee or Free: Free.

Aviation Safety Web Pages

http://web.inter.nl.net/users/H.Ranter

RATING
✈ ✈ ✈ ✈

BRIEFING:
An award-winning, yet dark view into the world of airliner disasters.

Morbid and sometimes somber, Aviation Safety Web Pages may seem at first to cater only to the accident curious. Those who are drawn to the disaster scene with fascination will certainly get their fill of fatality statistics and gory details. If you reach a little further, though, you'll notice this airliner accidents site employs descriptive accident data—the kind that may lend a hand with your own accident prevention preparation.

Religiously updated, the site contains a wealth of airliner accident info found in features like: The Aircraft Accident Database, Statistics, Accident Reports, Eyewitness Accounts of Accidents, Cockpit Voice Recorder ("Black Box") Transcripts, and a huge list of QuickTime movies, videos, and pics. Lists of accidents by year reveal endless summary tables of data. Event details like aircraft type, operator, and flight route combine with phase of accident, fatality counts, and remarks to paint a graphic accident picture. Links to additional articles, photos, and other reference sources round out your research options.

Have an unfulfilled curiosity about airline disasters? New fatal airliner accidents are added within one or two days. Sign up to receive free, e-mail "digests" to keep you apprised. Site update notification is also available for the asking.

Fee or Free: Free.

Jane's Information Group

http://www.janes.com

RATING

BRIEFING:
Almost reaching Web perfection overload, Jane's showers you with defense, aerospace, and transportation resources.

Fee or Free: Free to view, but a myriad of informational products are just waiting to be ordered.

For those who may not know, "Jane's" refers to the legendary English pioneer, John Frederick Thomas Jane, who began the widely regarded source for defense, aerospace, and transportation information. How appropriate that the world renowned resources of Jane's have found their way onto the cyber-info highway. Jane's boldly celebrates its 100 years of information solutions with its stunning example of Web design excellence.

Whether you choose to view the wondrous graphical imagery and organizational delight or opt for the speedy text version, you might catch yourself muttering praise as you click. Fascinating information and thorough descriptions are everywhere. Find out more about Jane's with The Product Catalog, Who's Jane?, or Editor's Notes. Take a ride into What's Hot with Photo of the Week, Interview of the Week, News Briefs, New Products, and more. Then, with your curiosity piqued, move into the information gallery. Here you'll uncover regional assessments, defense glossary (over 20,000 defense-related acronyms and abbreviations), and resource links (categorized by government/intelligence, defense/aerospace, space, and transportation).

Don't misunderstand. There's a lot of publication selling going on here. But, from the people that brought you the bible of the aviation industry, *Jane's All the World's Aircraft*, you can expect to be more than satisfied with brilliant content.

Ultralight Flyer Online

http://ul-flyer.com

RATING

BRIEFING:
Get airborne with this ultra-organized, ultra-balanced, ultra-designed, and ultra-bookmarkable ultralight site.

Boldly buzzing within aviation's online airspace, Ultralight Flyer Online gets in the cyber-pattern too with its own "ultra-site." Its award-winning offerings flutter around freely with simple euphoria. While layout, organization, and esthetics are thankfully in line with some of the better aviation sites, my favorite facet is informative link summaries. Mostly, I've noticed Web sites enjoy launching you into unexpected territory, under the hood with no instruments. I'm sure you've been on the frustration end of "follow-me" site topics that never rise above ground effect. Ultralight Flyer Online begins to erase the link maze stereotype with handy menu descriptions before you click—even submenus give you a nice preflight before departure.

The thoughtful site framework doesn't fly solo however. The site's notable organization teams up with a nice collection of content. Menu items include: Prop Wash—introducing you to an online digest for ultralight and sport aviation enthusiasts; Ultralight News—in newsgroup or newsletter format; Ultralight Reference Library—source for flying clubs, aircraft showcase, flight instruction, regulations, books, and videos; and UltraLinks Web Directory—pointing you toward other cool ultra-offerings with a handy search tool.

Fee or Free: Free.

FlightLine

http://www.flightlinemag.com

RATING

BRIEFING:
This "quasi-zine" (they've steered away from the traditional magazine boundaries) serves up flying features and news.

When an aviation information site offers an update status listed in minutes, I'm immediately interested. Combine continuous updates with fascinating feature stories, and I'm hooked. FlightLine has the "zine" ingredients most have come to expect, but their flair for variety is unique.

Sort through military aviation, model aviation, and flight simulator info. Read the latest news, topical features worldwide, or delve into some personal stories. Whatever your fancy, a constant main menu flies right seat with you—always ready with new topic coordinates. Expect to uncover the latest in global airshow news, be bombarded with a great collection of photos in the gallery, and find pointers to some highly regarded links (new ones come complete with descriptions). Finally, this award-winning info gem rounds out your cyber options with a download area (flight simulator & game stuff), computer hardware info, and ongoing messages from the Editor's Desk.

Other innovations you may appreciate include the latest news from Armed Forces Radio, free classified ad listings for 30 days, and a thoughtful French version.

Fee or Free: Free.

Airfax

http://www.airfax.com

RATING
++

BRIEFING:
Unlatch your tray table and prepare for a virtual smorgasbord of in-flight entertainment tidbits and juicy rumors.

Just the facts on in-flight entertainment are now only a click away. Airfax Online skims the commercial carrier scene for the latest in-flight industry insights and conveniently serves up the behind-the-scenes info. Onboard amenities, seat back video screens, air phones, meal services, and current carrier stats seem to be among the hottest topics. Aircraft purchasing contracts and passenger marketing complete the update.

This proficiently pleasant site eases you into recline mode with well-designed layout of five menu categories and interspersed current events. Browse the summarized feature story or click into the full text. Press the topic icons for: a summarized list of industry news & special reports; *Airfax IFE Newsletter* subscription and sample issue; company services; references for surveys and statistics; various links; and a prompt for site feedback.

Do keep in mind the actual *Airfax IFE Newsletter* is a subscriber service. Although you'll be introduced to an online sample, a fax or e-mail version will arrive twice monthly with your subscription. If you're in the business and rely on what the insiders have to say, this is your ticket. Have a nice flight.

Fee or Free: Free for some news and info—fee for newsletter subscription.

US Aviator

http://www.us-aviator.com

RATING

BRIEFING:
Current aviation news that won't waste your time.

Expanding upon their print version magazine, USA's Netflight Central is your Dan Rather for aviation news. Current event "news flashes" as well as weekly e-mail updates keep you flying-wise. Special features key in on events and guide you through a wide assortment of aviation links: The Proposed Airman's Bill of Rights, Aerobatics and Airshows, Airports, A/C Manufacturers, Aircraft Rentals, Airlines/Commercial, Balloons and Blimps, Colleges and University Programs, Civil Air Patrol, Flight Planning, Fly-ins, Insurance and Finance, Aviation Humor and Jokes, Legal Issues, NASA, Skydiving, Software, Ultralights, and more.

Browse the mall or peruse back issues. It's clean, uncluttered, and written well.

Fee or Free: Free, but leave your e-mail for weekly updates.

AVWeb

http://www.avweb.com

RATING

BRIEFING:
Free membership after completing a survey entitles you to competent aviation related journalism.

A daily info resource. Yes, daily. One of the best, most competent online aviation pubs around. Get your news each day, or browse in summary form with the weekly e-mail edition—AV Flash. The managing group are seasoned pros—writers, editors, and publishers with years of experience. You'll recognize names of well known regular contributors—the best in aviation journalism.

Selections include: NewsWire, ATIS, Safety, Airmanship, System, Avionics, Places, Products, Shopping, Classified, Brainteasers, Weather, Sites, and more. Also, dip into an organized database of: FAA Aircraft, Airman and Mechanic Registries, the Medical Examiner and Repair Station Lists, U.S. Airman Directory, the FARs and others.

You'll find a host of site navigation features, including menus everywhere you go, clickable topic icons, and convenient descriptions. The "New" section is a great personalized version of the table of contents. It shows you exactly which sections, articles, and features have been added or changed since your last visit.

I've been an avid subscriber for well over a year and a half now, and let's just say it's tops on my bookmark list.

Fee or Free: Free, but sign up for all the good stuff.

Air Chronicles

http://www.cdsar.af.mil/air-chronicles.html

RATING

BRIEFING:
Brought to you by the folks at the Airpower Journal, Air Chronicles gives you online insight into today's modern Air Force.

Air University Press, publishers of Airpower Journal, have opened the door to this interactive Air Force info resource. It's about Air Force doctrines, strategy and policy, roles and missions, military reform, personnel training, and more. Jump into a discussion group, or read what others have to say.

A host of categories include: Airpower Journal, Contributor's Corner, Airpower Journal International, Current Issues, and Book Reviews. My favorite, Contrails, gives you a mixed bag of aviation stuff, like: news, weather, tools, references, government sites, symposia, professional development, fiction, and professional journals.

Military authors, civilian scholars, and people like you and me create this heads-up, intelligent review of our modern Air Force.

Fee or Free: Free.

Aviation Digest

http://www.avdigest.com

RATING

BRIEFING:
Mainly a look into flying clubs and flight museums with a couple of fun additions. One hundred dollar hamburger anyone?

Do you enjoy bed and breakfasts, flying into Canada, or just soaring in Houston? You'll find a club that meets your needs. Search throughout the U.S. or Canada for a new group of buddies who share your interests aloft. Specifically, clubs you may find interesting include: The B&B Fly-Inn Club, The International F104 Society, South Central Section (SCS) 99s, Southwest Flying Club, TSS Flying Club, and Wings Over Canada. Still mesmerized by the abundance of info? Just type keywords into a handy site search—you'll find your way.

Scan through museums, helicopter services, the Advanced Maneuvering Program, links, and classified ads. You'll even discover sumptuous fly-in-and-dine hot spots nationwide—brought to you by the One Hundred Dollar Hamburger people.

Fee or Free: Free.

Aviation From Pilot—the UK GA Magazine

http://www.hiway.co.uk/pilot

RATING
✦✦✦

BRIEFING:
An OK UK online "zine" (representing its printed sister) with a fantastic A-to-Z list that sheds light on acronyms, abbreviations, and jargon.

Revolving upon the general aviation scene in the UK for over 30 years, *Pilot* magazine's printed version has made its mark in over 27 countries. Now with the online edition stretching out worldwide, those interested in UK aviation stuff need only check in here.

Not a UK flyer? Well, don't go just yet. There's a few things here for everyone: book reviews, CD ROMs, videos, past *Pilot* articles on instruction, flight tests, touring articles, and yes, a great reference resource for aviation acronyms, abbreviations, and jargon. Just click on a searchable alpha letter and get a succinct aircraft definition. From Accelerate-Stop Distance to Zulu, it's excellent for beginners and others puzzled by aviation jargon.

Fee or Free: Free.

Aviation & Aerospace
by McGraw-Hill

http://www.mcgraw-hill.com/aviation/aviation.htm

RATING
++

BRIEFING:
Aviation & Aerospace gives you a peek into your favorite McGraw-Hill aviation publications—featuring advertising and subscription info.

For leading edge insight into aviation, turn to the pages of McGraw-Hill's popular printed resources. If you're like me, you just can't get enough aviation info—whether online or in print. This aviation index page gives you at-a-glance briefings on the industry's leading magazines and newsletters. Perfect for planning your upcoming advertising schedules or just getting more details before subscribing, this site provides you with contacts, phone numbers, related products, leading advertisers, and demographic data.

Take a sneak peek into *Aviation Week & Space Technology*; Aviation Week Group Newsletters (seven publications ranging from business aviation to airports); Aviation/Aerospace Online (news and data available by subscription); *A/C Flyer* (corporate aircraft sales); *Business and Commercial Aviation*; and the *World Aviation Directory*.

Fee or Free: Free to peruse, but various fees apply when subscribing to magazines.

Inflight USA Magazine Online

http://www.inflightusa.com

RATING

BRIEFING:

Here's an in-flight magazine you'd actually want to take with you. Lucky for you this handy little "e-zine" is on the Web—no need to waste an airline ticket to get it!

Stow and lock your tray table. Return your seatback to its upright position. Give your flight attendant all that garbage you're holding. And, most importantly, don't forget to snag that in-flight magazine! Okay, truth is, they're only good for a captive audience. But, this online magazine (also existing in print) piques your in-flight interest a little better.

Soaring with juicy, current aviation info (stemming from each monthly printed version), you'll tap into a variety of hot topics. Dive into: politics, aircraft production, general aviation, legislation, etc. Or, get info on a variety of special topics like: air racing, aircraft reviews, Aircraft Owners and Pilots Association, historical aviation, new product news, safety issues, and more. During your online stopover, you'll also be invited to get sidetracked with a nice selection of stable links (lots of NASA stuff).

Oh, and if you just can't fly without the printed version, you can simply e-mail a subscription request. They make it so easy.

Fee or Free: Free for online viewing, fee for printed magazine subscription.

Internet Business Air News

http://www.bizjet.com/iban/default.html

RATING

BRIEFING:
A down-to-business air news site—efficiently designed to foil time wasters.

Enough flying fun, let's get down to business. With Internet Business Air News you immediately get the impression that dedicated, responsible adults are pulling the strings on this site. It's mostly text, organized well, with a few photos (I'm a fan of bandwidth efficiency).

Specifically focusing on daily business aviation news, the opening page has all the current events you can handle—no clicking necessary. All newsworthy articles load initially by default (from most current to least current). Then, should you need to scan back issues with past articles, just click to access.

After you've caught up on today's headlines, you'll find: a great directory of maintenance/FBOs/handlers (mainly Europe-based), directories, other sites, Virtual Air Show, and *European Business Air News* (printed magazine subscription info). Or, if you're a plane shopper, the Aircraft Market makes use of a pull-down list of aircraft manufacturers. All are easy and convenient.

Fee or Free: Free.

The Professional Aviator

http://www.iquest.net/propilot

RATING

BRIEFING:
A great up-and-coming resource site—evolving even as you read.

Keep your eye (and bookmark) on the Professional Aviator—it'll be a winner in time. At the time of review, info under most headings was a little sparse. However, info and articles that were in place hints at the professional flavor this site's creator means to achieve.

Touted as "the resource place for pilots," the Professional Aviator is obviously working hard to give you a nice cross section of aviation stuff: In the Cockpit focuses on flight safety issues; the Weather Page taps into weather, flight planning, and DUAT services; Rusty's Place offers words of encouragement, tips, and miscellaneous pilotage info; and Aviation Stories and Jokes brings you lighthearted chuckles. A smattering of worthy aviation links includes: Aviation Airworthiness Alerts, Advances in Medical, and FAA-Related Sites. There's also handy shopping links to sites like Jeppeson, Wings America, John Tackett Aviation Art, and Air Source One.

The Professional Aviator site is good now and getting better.

Fee or Free: Free.

Journal of Air Transportation World Wide (JATWW)

http://cid.unomaha.edu/~unoai/journal.html

RATING
+ + +

BRIEFING:
A scholarly online endeavor which will pique interest among enthusiasts.

Before sifting through this site, it may be appropriate to fix your bow tie, position the specs on your nose just right, and have aviation theory volumes at the ready. Before intimidation sets in, browse through the offerings—you *will* learn something. The Journal's goal is to eventually become "the preeminent scholarly journal in the aeronautical aspects of transportation." Lofty? Yes, but the JATWW will offer an online sounding board for peer-reviewed articles in all areas of aviation and space transportation, research, policy, theory, practice, and issues. After article review, approved manuscripts will circulate via list server free to all subscribers.

At year end, bound volumes, including all accepted manuscripts, will be available for sale and library reference. If you're not interested in being a part of the free global distribution, you can still visit the site and review any articles of interest. Topics include: aviation administration, management, economics education, technology and science, aviation/aerospace psychology, human factors, safety, human resources, avionics, computing and simulation, airports and air traffic control, and many other broad categories.

Fee or Free: Free, but you'll need to add your e-mail address to their list—follow the specific instructions carefully.

The Aviator's Hangar

http://www.img.pair.com/aviators

RATING
✈✈✈

BRIEFING:
Being grounded and hangar-bound can be entertaining with these well-written jewels.

The Aviator's Hangar eases that dull throbbing sensation resulting from flying withdrawals. When you're grounded and banished to the hangar by weather or repair, get a fulfilling fix with a cyber-stopover here.

Different from an aviation directory, this site provides a nice collection of entertaining and informative articles. Rags to Rivets gives you excellent tech tips on covering and refurbishing via a series of articles. Events and Shows lists popular events near you with useful descriptions attached. Stories contains tips, jokes, and editorials written by aviation writers and just ordinary folks. The Student Forum features a hodgepodge of questions, answers, and general online instruction from a seasoned pro. On those days of extra long hangar lounging, visit: The Link Library, Aviation Mall, Weather, Great Destinations, Bygone Aviation, Free Classifieds, Art Gallery, Live Chat, and more.

Although you'll be sidestepping a few typos, the articles and other features have those two key ingredients that shout AWARD WINNER: efficient and useful.

Fee or Free: Free.

The Avion Online Newspaper

http://avion.db.erau.edu

RATING

BRIEFING:
An online university "zine" that spreads its wings and intrigues more than just its fellow students.

No stranger to aviation news, the talented folks at Embry-Riddle Aeronautical University continue journalism excellence with this award-winning online newspaper. Mostly written for students, The Avion Online harbors a wealth of fascinating info for every flying enthusiast. You'll uncover the most current events in aeronautics. Find headline news makers on the front page and enjoy a host of University stuff. Categories include: Campus News, Metro News, Student Organizations, Space Technology, Data Technology, Diversions, Opinions, Sports, and Comics.

Technically organized with perfection in mind, clickable action buttons and eleven illustrated icon boxes make navigation simple. An online search engine requires only a concept or keyword for subject lookups.

Billed as "for students by students," The Avion gives us all an educated peek into aviation.

Fee or Free: Free.

Air & Space Home Page

http://www.airspacemag.com

RATING
✢ ✢ ✢ ✢

BRIEFING:

An online "zine" sibling to the printed version with its own distinct personality.

As your collection of unread aviation mags continues to pile up (due to too much flying time), stroll on over to the keyboard and tap into this award-winning online magazine. Sure, you'll be encouraged to subscribe to the well-respected print version, but if you're interested, the online way is easy.

Just here to browse? Well, stay awhile—it's worth it. The Air & Space site gives you a peek into the current hard copy issue and provides many additional online articles and features. For quick text-based navigation, the table of contents lists everything, including pages not referenced in the home page. Capitalizing on a history of journalistic excellence, Air & Space continues the tradition here with fantastic current events, interviews, reviews and previews, marketplace, associations, and more. You'll even find a few cute gems like embedded QuickTime movies, time/location-sensitive welcome info, and page animation.

After you've sampled and read through Air & Space's offerings try the visitor contact InfoBase. It's a handy all-purpose bulletin board and contact center. Get in touch with other aerospace enthusiasts with this searchable resource—it's excellent!

Fee or Free: Free (fee for magazine subscription).

Bookmarkable Listings

The Controller
http://www.thecontroller.com
Sneak peek into *Controller* magazine with for-sale classified and broker listings.

FsFan BBS
http://www.xs4all.nl/~fsfanbbs
Flight simulator information, links, and FsFan NET membership.

UK Airshows Review
http://www.uk-airshows.demon.co.uk
Personal UK airshow reviews, complete with future show information and past pictorials.

Aeronautx
http://www.aeronautx.com
Aviation industry news and information resources.

GPS World Online
http://www.gpsworld.com
Online extension of the printed magazine, *GPS World*, offers industry news and GPS-related resources.

Aerocrafter
http://www.baicorp.com/aerocrafter
Online complement to the printed magazine offers homebuilt aircraft info and sources.

Skydive!
http://www.afn.org/skydive
Award-winning skydiving archive, complete with equipment, training, and organizational information.

Aviation Disasters
http://www.gate.net/~avcrash
In-depth look into fatal commercial aircraft accidents with the hope of increased understanding and safety.

Flight Forum
http://webusers.anet-dfw.com/~toddc/flight.htm
Newsletter subscription site for aviation-related safety issues and information.

United Space Alliance
http://www.unitedspacealliance.com
Online space info center features Space Shuttle news, virtual facilities tours, and space-related resources.

National Championship Air Races
http://www.airrace.org/index.html
Official news and information site of the Reno Air Racing Association and the National Championship Air Races.

Aviation Parts, Supplies, & Aircraft

Aviation Industry Resource (AIR)

http://www.air-air.com

RATING

BRIEFING:
A versatile, searchable link library with handy aircraft-for-sale resource.

Just as most modern libraries provide computerized searching to quickly locate your needed advice, Aviation Industry Resource has online volumes at the ready. Free to everyone, this international aviation registry and listing service quickly finds links, aircraft, and navaid info based upon your specifications.

For me, quick and easy page navigation is "resourceful" enough never mind AIR's simple searching forms. Jump into the link registry, pick your category, click a location, and type a keyword or two. You'll be thrilled with response time and the sheer volume of thousands of links. Adding your own Web site address (if you have one) is completely free and easy. The aircraft-for-sale listings work similarly. Simply pick a desired make of aircraft and get a corresponding list. Additions to the for-sale list are also free. I didn't uncover much in the listings area as of review time, but the collection seems to be growing—especially with no fee.

Other clickable site topics include: Sponsorship, Web Sites, Flight Planning (a few helpful links), and the fairly popular Bulletin Board.

Fee or Free: Free, even to list a for-sale-aircraft or add your Web site address.

Aerosearch

http://www.aerosearch.com

RATING
++

BRIEFING:

Self-dubbed, "the world's first free and searchable aviation database," you'll find your tools and parts here.

Yes, the icons are funky and overall site esthetics leave you wondering who was flying left seat on the layout mission. But, before you make a mad dash for your favorite bookmark, dig a little deeper into the heart of the content. Disappointment completely subsides when you engage in a search for aviation parts and tools.

Touted as the world's first free & searchable aviation database, Aerosearch does require new user registration (free) before you're underway. After login and password credentials are checked, you'll enter part numbers, alternate part numbers, or the national stock number. There's also an option to include up to three different search words. Then, click "start search" for your comprehensive list of suppliers.

Yes, the AeroPort section has more in the way of searches and info. New Equipment Manufacturers, FBOs, Aircraft for Sale, Repair Stations, Engine Shops, Airports Around the World, Flight Schools, Avionics, and Rental/Charter topics are all a bit sparse (as of review time).

My advice? Just stick to parts and tools searching for now.

Fee or Free: Free.

VisionAire Corporation

http://www.visionaire.com

RATING
✈ ✈

BRIEFING:
Peek in for visionary vistas of succinct corporate information.

Sure you've got your Boeings of the cyber-world spewing out online magic with infinite resources standing by. But, while obviously not the manufacturing powerhouse of Boeing, the VisionAire Corporation demonstrates its Vantage aircraft and other company topics with skill and purpose.

Its compact, yet finely tuned Web offerings give you concise insight into six main topics. Click on What's New to stay abreast of site changes—you can even register to receive automatic page updates. About VisionAire moves you into: The VisionAire Story, Meeting the Team, Company Locations, and Our Mission & Values. The Vantage Link supplies info regarding why you should consider a Vantage, Airplane Specifications, Aircraft Comparisons (under construction at review time), Prices & Terms, Partners & Suppliers, and Photo Gallery. You'll also get more insight with VisionAire News & Events, as well as the latest employment opportunities.

Fee or Free: Free.

Aviation Café

http://www.avcafe.com

RATING

BRIEFING:
Order your desired aviation tidbits with the Aviation Café. From specials to menu items, it's all good.

Sumptuous content entrees team up with tasteful ambiance to stir a pilot's online passions. Although reservations aren't required, the 24-hour Aviation Café sports a flair for atmosphere and tempting offerings. From the moment you check in, you realize this "ain't no airport dive eatery." Stylish icons, roomy page layouts, and easy-to-use navigation menus come with your content entrees. Tasty descriptions tell you a little something before you click into a hearty topic, and easily understood directions can be found throughout.

List your airplane for free with an easy-to-use online form (there's a bunch as of review time). If you're more on the aircraft shopping side, scan through the listings—most come with pics. Search by: single engine, amphibian, homebuilt, list by price, multiengine, helicopter, turbine, and those listed in last 30 days. Another savory menu item to be considered is the Café Quiz—a daily interactive way to stay current. Test your knowledge of FARs and other basic aviation topics.

Today's special (as of review time): free aircraft-for-sale listings. They won't be free forever, so get your listing in while they're hot.

Fee or Free: Free.

AvShop.Net

http://www.avshop.net

RATING

BRIEFING:
No lines. No waiting. Step up and satisfy your pilot supply needs with online convenience.

Even though it's graphically oriented, AvShop.Net's online catalog loads fast. The layout people (who are obviously talented) must have meshed perfectly with the site mechanics people, because the final result exemplifies teamwork.

Once the initial welcome page instantly appears, you begin to sense that the interior will be just as fancy as the exterior. Click to enter the catalog and you won't be disappointed. Because, lurking under the cowl of this aerodynamic machine beats the heart of a full-blown aviation supplies catalog. Mostly magnifying its software selections, AvShop.Net still provides a wider selection of necessities than you're led to believe. A nicely presented main menu moves you quickly into stuff like: books and study guides, aviation videos, cockpit and aircraft accessories, pilot supplies, and a healthy dose of software specialties.

The site's beauty and brains also rise to the occasion during the "checkout." Once you've filled your basket with goodies, a well-organized system of ordering ensues. Review your basket contents, change or remove items, and proceed to either an electronic or off-line "checkout." It's easy and convenient.

Fee or Free: Free, unless you're tempted into ordering.

Spinners Pilot Shop

http://www.spinnerspilotshop.com

RATING

BRIEFING:
As online supply shops go, Spinners tops the list with revolutionary completeness.

Few pilot supply sites handle the complete online ordering system flawlessly. In fact, only three come to mind, with Spinners propelling its way onto my list. Layout and design work well together with little waiting time. Colorful menu buttons in the left margin are always a click away. Online transactions are given secure routing. But, the best part revolves around Spinners' huge inventory. It's not the online mirage that most aviation supply sites throw together— just read some of their customer testimonials.

Without even touching the subcategory list, I'll give you a taste with: headsets, Jeppesen, pilot supplies, books, logbooks, software, flight computer, flight bags, videos, training aides, kneeboards, flashlights, GPS, intercoms, handhelds, plane supplies, and more!

Shopping is relatively painless with the standard "shopping cart approach." Simply view product photos, read descriptions, review prices, and add items to your "cart." Then, check out online or offline. Simple instructions guide you through either method.

Get online, grab a mouse, find your favorite plastic card, and spin your way into a buying frenzy. It's easy to do.

Fee or Free: Free.

WWW.Plane-World.com

http://www.plane-world.com

RATING

BRIEFING:
Kick the tires, rattle the flaps, and check for prop dings online with Plane-World.

All review criteria aside for a moment, the real test for aircraft-for-sale sites comes down to volume of listings. Selection is key, and to be frank, most aircraft-for-sale sites fall dramatically short in the listings area. Worse still, fly-by-night cyber-brokers tend to throw up visually forgettable Web tangles, expecting you to find your way through their meager offerings.

Rising above the scattered masses with Vx climb, Plane-World blasts onto the for-sale scene like a trusty Cessna 172 gone turbine. Appropriately simple menus and well-designed layout clear your way into a huge variety of classified listings. Following a left-margin menu that changes into submenus with each topic, page reckoning is of the 100-mile visibility variety. Start with For-sale, Wanted, or Latest Listings. Then, try the handy search engine, place your own free classifieds, or scan The Miscellany (employment, engines, propellers, radios, and electronics).

Although picture-less, the listings couldn't be more efficient and informative. The lack of pics may put off some, but that's why the site's so fast. It's a trade-off I'll take gladly.

Fee or Free: Free to browse and to add your own classifieds.

Aeroprice

http://www.aeroprice.com

RATING
++

BRIEFING:
Holding your hand through buying or selling, this site's cool features make for a worthwhile stopover.

Buying and selling aircraft. In the lifetime of most serious aviators, the murky abyss of either endeavor may be riddled with uncertainty—especially the first purchase or sale. But, breaking through the low lying fog of confusion, Aeroprice offers a progressive taxi toward understanding.

Though limited on any visual wizardry, Aeroprice provides some handy fee-related services, free tips & trends, and a gentle push toward additional resources. No, I'm not navigating under the hood. I realize that this for-sale site is subtly maneuvering toward cyber-sales of its QuickQuote online pricing and appraisal software. But, it's certainly worthy of a flyby if you're shopping or selling.

After entering info into a thorough online questionnaire, you'll be introduced to an excellent aircraft pricing analysis. Look for great insight into pricing adjustments based upon the average retail cost for your selected aircraft. Items analyzed include: airframe, engine, avionics, additional equipment, interior, exterior, and damage history.

Fee or Free: Free with some fee-related services.

McDonnell Douglas (MD)

http://www.mdc.com

RATING

BRIEFING:
This big corporate player weaves a Web of company info and fun gadgetry to satisfy all aviation surfers.

Among the mega-sites that spin Webs of visual wonder, McDonnell Douglas ranks right up there, where oxygen is needed for extended periods of time. If you're a mega-site admirer, odds are you'll virtually hyperventilate at this site's displays, gizmos, content, and gadgetry.

A surprisingly quick-to-load pictorial collage represents your introductory menu—complete with current McDonnell Douglas stock price and text-based hyperlinks. As you would expect, company info oozes out of every link. Learn about MD's profile, history, vision, community and environment development, annual report, speeches, and more promotional matter. True MD junkies might even want to delve deeper into its customers, product support, supplier management, airport technologies, technical services, and more. The more interesting facets offered are: employment opportunities, MD-sponsored Blue Angel's page, and a look into the company's space involvement.

If you're not into the corporate-type information, skip over to the fun frivolities. Download the colorful screensaver, click through the multitudes of photos in the gallery, or browse through a great series of video files.

Fee or Free: Free.

Bombardier Aerospace Group

http://www.aerospace.bombardier.com

RATING

BRIEFING:

Jet shoppers or just dreamers will find the ultimate in online perfection with Bombardier.

If you were searching for that flawless corporate aerospace site against which to compare others, Bombardier Aerospace Group's Web presence is the model. Organization and page designs are among the cutting edge variety. Photo gallery images are professionally striking. And, page navigation is effortless. Though a fully functional and informative gem, the site is simply a standard-setting masterpiece.

Okay, so maybe you're not in the market for a Canadair Regional Jet or a Challenger, but most enthusiasts will agree that the company's line of aircraft are worth an appreciative peek. Luxurious business jets like Global Express, Challenger, Canadair, and Learjet grace the company's online pages with specs and pics. Similarly, regional aircraft info for the Canadair Regional Jet and the de Havilland Dash 8 series are at the ready.

Those more interested in employment rather than jet shopping will appreciate their extensive online personnel department (originating in Montreal). Detailed job offerings are categorized into Administration, Customer Support, Information Systems, Engineering and Manufacturing.

Fee or Free: Free.

Northrop Grumman Corporation

http://www.northgrum.com

RATING
✈✈✈

BRIEFING:

Zero in on combat and weapons technology info with Northrop Grumman.

When you're talking about combat aircraft, precision weapons, or defense electronics, technologies developed by Northrop Grumman will usually wind its way into the conversation. When the discussion expands into online resources, Northrop Grumman's Web site must be in the mix. The company-wide facts, news, and knowledge flow endlessly within its online offerings. I suppose you'd expect a masterful site from a company who plays a major role in many of the world's most advanced weapons systems and technologies. After all, they're no strangers to design and systems integration.

The main menu guides you through the exhaustive company tour with broad categories of: What We Do, What's New, Executive Officers, *Review* Magazine, Photo Gallery, Corporate Directory, Career Opportunities, and more. Background info, product fact sheets, and key personnel biographies are well-written and interesting.

Although you might be stifling a yawn, ready to turn the page, hold tight. There's fun stuff here too. I particularly enjoy the great images in the video gallery of: B2 Missions and Testing, Various Aircraft on Display, Manufacturing, The F/A-18 Hornet in Action, and more. Quite frankly, the videos alone are reason enough to pay a visit.

Fee or Free: Free.

Lockheed Martin Corporation

http://www.lmco.com

RATING
+ + +

BRIEFING:
Lockheed Martin is light years ahead of its time in understanding how to appeal to aviation enthusiasts with the Web.

With ornate visual esthetics riding shotgun, Lockheed Martin obviously sat the content people in the left seat for this highly informative site. Early on during this online flight you're inundated with text-based introductions into current news highlights and noteworthy events. Lists of topics literally fill the pages under category headings of: Highlights, Upcoming Events, Aeronautics, New Business, Electronics, Launch Information, and more.

You're looking for examples, aren't you? Remembering that these will change after review time, you'll be browsing articles like: "New U.S. Air Force F-16s Will Have Color Displays and Other Advanced Systems;" "Manned Space Systems to Produce Tanks for Reusable Launch Vehicle;" and "Lockheed Martin Completes Initial Design Review for Its Joint Strike Fighter Program."

Once you've muddled through lengthy topics and summarized tidbits, do check into the library—the resources are endless. There's a Lockheed Martin photo archive, video library, Lockheed Martin Today (online periodical edition), and *Code One* (the company's Tactical Aircraft Systems quarterly magazine).

Fee or Free: Free.

Microsoft Flight Simulator
http://www.microsoft.com/games/fsim

RATING
+ + +

BRIEFING:
Get the real briefing straight from the source—before you fire up those flight simulator engines.

The Flight Simulator site combines a couple of magical components inherent in all award-winning sites: valuable online information and easy communication to a huge audience (I bet you know a Microsoft Flight Simulator user).

Sure there's software promo stuff everywhere. Ad banners and purchasing pages abound. But, looking a little further into the mix, Flight Simulator users will strike virtual gold. The latest product information and downloads can be found in News. You'll find lively discussions, topical chats, and tips in the MSN Flight Simulator Forum. Performance specs, flight scenery enhancement products, and current Flight Simulator news are also handy for cyber-flyers. And, for those not yet into the Microsoft Flight Simulator scene, a convenient demo movie gives you a free sneak peek at this wondrously real simulator.

Whatever your simulator readiness, all Microsoft Flight Simulator users need a quick refresher here. And, soon-to-be users? It's simply a perfect preflight walk-around.

Fee or Free: The info and tips are free. The software isn't.

The Aviation Online Network

http://www.airparts.com

RATING

BRIEFING:
An easy-to-use, partly subscriber based resource—great for finding parts.

The selections here aren't endless, effectively narrowing the confusion factor. Simply and conveniently at your fingertips are: Parts Search (for members), Follow Me for RFQs (an equipment buy & sell forum with free classifieds), Parts Search (demonstration), Show News, Links (broken out by icon-represented categories), and Weather (check the weather anywhere in the world).

If you're in the market for parts, you'll have good luck here. Some stuff is available without subscribing, but for parts you'll need to subscribe.

With a little extra ground time on your hands, you're invited to test your skills with the FAA Exam Man Question of the Day—you might even win a free video!

Fee or Free: Free, but you'll need to subscribe (name, address, e-mail, etc.)

Global Aviation Navigator

http://www.globalair.com

RATING

BRIEFING:
Good internal database makes searching for aviation stuff pleasantly efficient.

Don't be fooled. It's not *2001—A Space Odyssey*. Punching any of the multicolored buttons will guide you to info and company listings. Useful search engines narrow your focus—whether it's FBOs, airports, dealers, weather, insurance, FAA stuff, financial services, and more. Specifically, enthusiasts will find a satisfying collection of resources in: Reservations, Maintenance Services, Aircraft Charters, Pilot Services, Aircraft Parts, Suppliers and Manufacturers, Flight Planning, Avionics, and more.

Fee or Free: Free.

PC Aviator

http://www.pcaviator.com.au

RATING

BRIEFING:
Lots of selling going on here, but good products abound. Become introduced to some quality flight simulator stuff—great for power off stalls in your pj's.

Find flight simulation hardware and software on this site Down Under (Melbourne, Australia). Approved for serious training purposes or just frivolous fun, PC Aviator's flight simulator products give you lofty experiences straight from your computer. To enter, click on the runway.

Although a hard sell for their simulator stuff, the site offers a limited selection of other flight simulator links, a great list of worldwide aviation links, and downloadable software. Other clickable areas include The Latest Headlines (site news and miscellaneous flight simulator links) and *Computer Pilot Magazine*.

If you're interested in the PC Aviator full color catalog, just sign up here. This printed product catalog is full of screen shots and simulator product info. By adding your name to the mailing list, you'll regularly receive a copy of the catalog.

Fee or Free:
Free—sign up for the catalog if you're a "sim" buff.

Aircraft Shopper Online (ASO)

http://aso.solid.com

RATING

BRIEFING:

It's an award-winning site for a reason. Unique searching capabilities smooth out online shopping turbulence.

Keeping the value of your time in mind (hey, you could be flying instead), ASO's pages are effortless and efficient. It's mostly text until you get to your selected destination. Once you plunge into over 800 aircraft-for-sale listings and narrow your search, you'll get descriptions and photos. Serious aircraft shoppers should skip directly to "PowerSearch" for excellent sorting and criteria setting. Set price and date ranges, limit the search to one aircraft make, or scan the entire listing. If you are (or will become) a frequent ASO shopper, these crafty Web developers have even included a clickable area for new ad additions and changes—showing only changes or added stuff for the last seven days.

Also handy are: Aircraft Dealers and Brokers; Aviation Links & Terms; Aircraft Partnerships; Aircraft Parts, Engines, Interiors & Repair; and the online assistance of Help & Tips.

Fee or Free: Free.

Air Source One

http://www.airsource1.com

RATING

BRIEFING:
Speed through this quick checkout line for student, corporate, military, and airline pilot supplies.

Grab an electronic shopping basket and stock up on your favorite aeronautical necessities with Air Source One. Simply put, you'll breeze through the quick check with no lines and everything you could possibly need.

This giant online pilot supply superstore expertly offers up aisles of products from which to choose: headsets, GPSs, transceivers, charts, other electronics, FAA test preps, flight bags, aviation books, apparel, gifts, necessities, and software. Conveniently complete your order by credit card (they assure secure credit card processing), phone, fax, or mail. Online ordering can be next-day delivered and includes an e-mail confirmation.

If your basket gets heavy and you just want to find a particular item in a hurry, choose Quick Search and simply type in your item. More minor site features include: weather, job opportunities, pilot forums, NOS latest editions table (great reference for chart dates and next editions) and other aviation sites.

Fee or Free: Free, unless you buy something!

Jeppesen

http://www.jeppesen.com

RATING
✈ ✈ ✈

BRIEFING:
Artfully organized electronic catalog offers useful pilot supplies.

With over 60 years of industry leadership (from a man who invented aviation charts), Jeppesen Sanderson once again captures the aviation world's attention with a visually captivating, organizationally brilliant Web companion. From the world's leading publisher of flight info (computer flight planning services, aviation services, and training systems), Jeppesen's site provides a current, online look into its offerings. Conveniently located among the company's history and profile info, you'll find the promotional core: the Jeppesen catalog.

While most "e-catalogs" weave the shopper through time-consuming pics and tangled disarray, you'll glide effortlessly through Jepp's: Manual Services, VFR Flight Information, Airport & FAR Reference Info, Airway Manual Accessories, GPS/NavData Services, Jeppesen FS-200 Instrument Flight Simulator System, Pilot Supplies, Maintenance Training Products, and CFI renewal program.

When ordering you have three options: 1) e-mail order info, 2) call the listed 800-number, or 3) fax the order. No credit card online ordering systems were available at the time of review.

Fee or Free: Free.

Wings Online

http://www.wingsonline.com

RATING
✈ ✈ ✈

BRIEFING:
Aviation shopping is made easy with this site—crammed with specs and pics of aircraft for sale, rent, or lease worldwide.

I'm always a fan of productive, visually appealing Web creations resulting from skill and a lot of elbow grease. The moment you grab the yoke here and finesse the controls, you'll also believe someone spent some late nights fine-tuning the many subtle nuances. This fantastic aircraft shopper resource gives tire kickers and eager buyers alike good info and many search choices. Sort by aircraft type, price, location, or the latest additions as of current date (marvelously efficient!). Search through related listings, such as: For Sale by Dealer, For Sale by Owner, Aircraft for Lease, Engine/Parts Wanted, and Aviation Real Estate. Once you narrow your search and actually tap into the seller's wares you'll be instantly informed with important specs: TT, STOH, SBOH, SMOH, registration, avionics, interior/exterior, price, contact, and more. Each listing also includes up to five pics showing off such areas as: exterior, panel, interior, etc.

The best part for you Net-savvy shoppers is you'll actually find loads of quality listings—not a smattering of local rejects. When you're ready to really shop, try on this site, it fits perfectly.

Fee or Free: Free.

Optima Publications

http://www.pilotsguide.com

RATING
✈ ✈ ✈

BRIEFING:
This Optima site deftly guides you through their popular printed airport guides.

Okay, so maybe I've slightly narrowed the geographic scope with Optima Publications. But, click through the Pilot's Guide Online and you'll see why they've become an award-winner. With a focus on California, Southwestern, and Northwestern Airports, the Optima publications offer vital and current airport info.

You'll find convenient and thorough descriptions of product offerings, including: Pilot's Guide to California Airports, Pilot's Guide to Southwestern Airports, Pilot's Guide to Northwestern Airports, Fun Places to Fly, Aeronautical Chart Subscription Service, and more. Click on any product category for instant prices and ordering info.

Scan through info about the Pilot's Guide (history and what it is), topics relating to current subscribers (revisions service, customer service, etc.), and a What's New section for frequent visitors.

It's a simple, well-organized sales pitch for some excellent pilot products.

Fee or Free: Free.

AirShow—Aviation Trading Network

http://www.airshow.net

RATING

BRIEFING:
This expertly arranged for-sale site combines thoughtful features and a wondrous assortment of quality aircraft.

You may, at first glance, give your own high rating to this site for its online design and user friendly layout. But, I urge you to discover the core of the AirShow's hidden talents. Just try looking up your favorite aircraft. While most aviation for-sale sites skimp by with only a few aircraft, the AirShow explodes with a huge variety of listings. It's a true resource for the buyer and seller. Clickable menu icons include: Aircraft for Sale, Dealers and Brokers, Manufacturer's Showcase, Find Buyer, Show Your Aircraft, Feedback (e-mail), and What's New.

The expertly developed searches within the huge database can be narrowed by: price range, aircraft type, special characteristics, and aircraft make/model. Serious plane hunters will rejoice at the brilliant What's New search—giving you only the latest additions since a user-specified date.

Not shopping, just selling? Well, you'll be equally impressed. All details for listing your unwanted flying machine are here—just click. Conveniently, you have a choice between an online ad form, or custom service (mail, fax, or e-mail your photos and info).

Fee or Free: Free. If you're interested in showing an aircraft, reasonable fees apply.

The Official Site for Learjet, Inc.

http://www.learjet.com

RATING
✢ ✢ ✢

BRIEFING:
Lear's self-promotion company site gives you history, sales info, and employment opportunities in a sleek, drag-free package.

Bet you didn't know that historically Lears have blasted beyond the bounds of aircraft production. You may be surprised to know they dabble in: electronic warfare simulation, aerial photography, airways calibration, air ambulance, target towing, radar training, fire control radar, electronic countermeasures, jammers, and 360-degree surveillance radar.

Learjet's official company Web site not only dips into history and fascinating company facts, but you can do a little shopping here as well (from Lear gift watches to Lear 60 aircraft). Not buying, just browsing? Well then, you've come to the right expertly-manicured place. Examine the Lear 31A, 45, and 60—complete with a series of downloadable pics (instrument panel to aft lavatory) as well as copious specs.

Those dazzled by the high flying Learjet may even want to consider employment opportunities. Organized by the NationJob Network, clickable job options abound. Simply click on a desired position. Job summaries include brief descriptions, experience/qualification requirements, compensation/benefits, and how to apply. If you're really Lear-curious, don't forget to peruse press releases and read about current Learjet breakthroughs.

Fee or Free: Free.

Raytheon Aircraft

http://www.raytheon.com/rac

RATING

BRIEFING:
Professionally prepared corporate look at Raytheon Aircraft.

Originating from the Raytheon company home pages, I invite you to skip directly to this nicely informative company page specifically devoted to Raytheon Aircraft. You'll quickly be enlightened about the company's broad product line. The organized format gives you efficient descriptions of the Hawker 1000, Hawker 800XP, Beechjet, Raytheon Premier I, regional airliners (Beech 1900D), business turboprops (King Air series), piston-powered aircraft (Bonanza's and Barons), and military aircraft.

Current company press releases (timely and updated regularly) give you tidbits like: "Raytheon Aircraft Delivers 5,000th Beech King Air," and "E-Systems Montek Division Merging with Raytheon Aircraft." News media have a special e-mail link called Corporate Communications. But, the rest of us may get more info (or actually talk to someone) by dialing the listed 800-number.

More clickable categories include: Key Business Areas, Shareholder Info, and Employment Opportunities. Still lost? A convenient site map makes finding a topic effortless.

Fee or Free: Free.

Boeing

http://www.boeing.com

RATING

BRIEFING:
The world's leading commercial airplane manufacturer blasts off with more Boeing brilliance.

Among the aircraft manufacturers vying for some of your Internet cyber-time, nothing Web-wide comes within a nautical mile of Boeing's online extravaganza. You could easily spend hours (even days) and not unearth every informative tidbit. Graphics, page navigation, pictures, facts, and surveys masterfully combine to create this interesting look into the world's leading manufacturer of commercial airplanes.

In addition to its leading manufacturing position, Boeing commands respect with its capabilities in, and informational Web pages relating to: space systems, helicopters, military airplanes, missile systems, information and electronic systems, and software products. Get an insider's peek into this jumbo company with a quick look at Boeing (at-a-glance info); Boeing in the News (news releases, financial reports, shareholder information, etc.); and Inside Boeing (products and services, history, gift store, tour information, and more).

Also wonderfully prepared are the pages relating to tours (complete with photos) and vast employment opportunities. The employment "area" is a grand affair with subjects relating to: college recruiting, internships, employment opportunities, advertising schedule, career fairs, pre-employment assessment, and Employment Center information.

Fee or Free: Free.

The New Piper Aircraft, Inc.

http://www.newpiper.com

RATING
✈ ✈

BRIEFING:

Piper product pitch pages captivate with nice pics and performance specs.

To get something out of the New Piper page you really don't have to own one or carry exclusive membership credentials from some elite Piper club. Even non-Piper junkies will enjoy browsing performance specs, pricing, and comparing equipment lists on the entire currently manufactured fleet.

Aircraft choices for further examination include: the Warrior III, Arrow, Seminole, Saratoga II HP, Archer III, Malibu Mirage, and the Seneca V. If the pages move you, a clickable map easily locates a dealer in your area. Or, if you're more an occasion flyer than a buyer, get the scoop on company tours with a factory tour info page, past Piper press releases, or browse pilot shop goodies.

Fee or Free: Free.

Bookmarkable Listings

Rockwell
http://www.cca.rockwell.com
Corporate information regarding Rockwell's avionics, communications, and navigation products.

Europa
http://www.europa-aviation.co.uk
Detailed descriptions and information regarding the Europa kit plane.

Avsupport Online
http://www.avsupport.com
Fee-oriented aviation parts searching.

U.S. Wings Aviation Mall
http://www.uswings.com
Manufacturer and distributor of aviation products.

WSDN Parts Locator
http://www.wsdn.com
Aircraft spare parts locator and repair/supplier database.

Internet Parts Locator System
http://www.ipls.com
Database of aircraft spare parts and repair capability for the commercial aviation industry.

007 Aircraft Classifieds Online
http://www.web-span.com/acsales
Fee-oriented, aircraft-for-sale listings.

Web Wings Ltd.
http://www.lainet.com/webwings/aviation.htm
Fee-oriented, aircraft-for-sale listings.

Airbus Industrie
http://www.airbus.com
Corporate overview includes news, photos, videos, and history of Airbus aircraft.

Airhead Pilot Shop
http://www.airhead.com/~catalog
Online catalog features a wide selection of pilot supplies.

Aircraft Suppliers Company
http://www.airsuppliers.com
Online catalog features a wide assortment of aircraft parts and accessories.

Aviation Entertainment

AVIATION ENTERTAINMENT

The Flight

http://www.gruner.com/flight

RATING

BRIEFING:
A fifty-year-old pilot and a fifty-year-old Cessna 195 trip the sky fantastic with this amazing personal account.

It's ironic that such an entertaining site devoted to hardy solo navigation offers slightly awkward Web site navigation. Just slip into the Table of Contents page, you'll be on course instantly.

Well-written chapters chronicle a 6,000 mile journey in a Cessna 195 radial engine beauty. Each online account uses powerful prose and striking photos to capture the essence of one man's fascinating skyward journey. His route includes: St. John's, Newfoundland (the most easterly tip of North America) and flying due southwest, over the Canadian maritime provinces, across the U.S., through northern Mexico, past the Tropic of Cancer to Cabo San Lucas, Mexico (the most southwesterly point of the continent). Scintillating chapters are titled: "Silent Giants of the Atlantic;" "Icing, Winds, and Silence;" "Flying the Gauges;" "Revolutionaries and Bandits;" "The Mythical Island of California;" "A Small, Dusty Airport;" and "The Pearl of Loreto."

This real account speaks of dark fjords, the brutal Sierra Madres, miles of oceans and deserts, and a mixed bag of weather. No, his wife didn't go.

Fee or Free: Free.

AVIATION ENTERTAINMENT

Dave English's Web Site
Great Aviation Quotes
http://www.skygod.com

RATING
++

BRIEFING:
Dave English and friends get lofty with words. Soak in this thought-provoking mishmash of visionary verbiage.

Yes, there's some in jest, others are rooted in seriousness. But, all the aviation quotes (more than you can imagine) do entertain. The hordes of quotes are conveniently broken down by category. Here's the list (take a breath first): Airports, Air Power, Balloons, Bums on Seats, Combat, Clichés, First Flights, High Flight, Humor, Last Words, Magic and Wonder of Flight, Maps and Charting, Miscellaneous Stuff, O'Hare ATC, Piloting, Poetry, Prediction (past and present), Press Corps, Safety, Space Flight, Women Fly, The B-17, The B-747, The Concorde, The DC-3, The F-117, The Harrier, and The P-38.

When you've finished absorbing these thoughts from meandering minds, click through Dave English's other site gems. Learn about the history of airport identifier codes. Plot a course through tremendous aeronautical chart resources. And, as always, stop by the links list—there's plenty.

Fee or Free: Free.

AVIATION ENTERTAINMENT

The Mile High Club

http://www.milehighclub.com

RATING

BRIEFING:
A scintillating site devoted to the erotic club that pilots, flight attendants, and daring airline passengers have been whispering about since early flight.

Tastefully exploiting the romantically inclined members of the Mile High Club, the Mile High Club Home Page boldly makes its online presence known. Mostly text-based, this mile high adventurer's site organizes a thoughtful mix of limited but appropriate graphics, page navigation perfection, and interesting stories.

Curious Mile High wanna-bes and members will find nonstop entertainment with this erotic fixation on skyward fantasies. Clickable sections include: All About the Mile High Club—membership requirements and rules/regulations; Mile High Adventures—info on one airline that specializes in this sort of thing; Mile High Club Store; and, of course, Tales of the Mile High Club—adult-oriented stories of those in pursuit of becoming a member; and more.

Okay, you're curious, right? Well, here's a taste of the tantalizing tales: "Full Aft Position," "Perks of the Job," "Commuter Fun," "Newlyweds," "Flight to Down Under," and more.

WARNING: Content of Mile High Club stories may contain language or situations not appropriate for minors.

Fee or Free: Free.

AVIATION ENTERTAINMENT

Plane Spotting

http://members.aol.com/planespot/main.htm

RATING

BRIEFING:

Enthusiasts' reviews are categorized and added to the worldwide plane spotting database. It's the right spot for spotting.

Whether you admit to it or not, all of us aviation types are apt to search the sky for a propeller's drone or a jet's roar as it goes by. For this reason it's always easy to spot the aviation enthusiast in a crowd.

Plane Spotting devotes itself to those who wouldn't pass up an opportunity to stop, watch, and revel in the wonder of flight. The site offers a plane spotting database of best aircraft viewing spots around the world. Entries that may be added by anyone are comprised of: directions to viewing spot, airport and identifier, city, type of aircraft frequenting the location, runway headings, frequencies, parking hints, and more.

Site navigation is simple with clickable links and short descriptions. A What's New section displays latest entries and info. Add an Entry provides an easy form to fill out. And, Aircraft Images does a great job of displaying some featured spottings without taxing your modem.

Fee or Free: Free.

200 Best Aviation Web Sites   **213**

AVIATION ENTERTAINMENT

The World of Aviation Poetry

http://www.gower.net/erst/airpoems.htm

RATING
✈ ✈ ✈

BRIEFING:

Don't let load time discourage your curiosity, these poems of aviation passion are timeless.

Hidden deep in this confusing cornucopia of online memorabilia, the heartwarming prose of Mr. E. Rowan S. Trimble summons you like an airport beacon. Hopefully, one reason you consult my collection of Web winners is to quickly arrive at such aviation gems as the World of Aviation Poetry.

The fantastic poetry is certainly worth waiting for—and you will be waiting for the disorganized jumble of a home page to load. Have a steaming mug of joe standing by, because efficient this site is not. The main reason you need to add this site to your favorites list? Heart-wrenching, thoughtful, inspiring, cleverly described words of wonder prevail. Some poems are short and powerful. Others linger through description and flowing recollection. My favorites are: "Vagabond Pilot," "The Final Cynosure of Fort Wolters," "Explaining Flight to My Son," "Broken Wings," "Echoes," and "Too Low." Yes, my favorites are many. I defy you to be more limiting.

Secondary reasons you'll enjoy your stay include: a series of great aircraft sounds in WAV format, Poetry in Aviation Pictures, Famous Flying Quotes, and much more.

Fee or Free: Free.

AVIATION ENTERTAINMENT

Dave, Carey and Ed's Lancair Super ES Kitplane Progress Page

http://www.edlevine.com/lancair/index.htm

RATING
╬ ╬

BRIEFING:
Follow along as three first-time kitplane builders take us for a ride on their Lancair learning curve.

By including Dave, Carey and Ed's Lancair Super ES Kitplane Progress Page among the top 200 aviation Web sites, you probably think I've pulled a few too many Gs and rendered myself unconscious. While those around me may disagree, I assure you I'm fully aware and rational as ever. You'll just need to ride right seat with me on this one and follow my lead.

First, a few warnings. You'll be horrified by site design. Site navigation simply means using your scroll bar. And, the long-winded text tends to run on through carefree punctuation. Sweeping all of the negatives under the rug, however, reveals an entertaining over-the-shoulder look at the slow progression of building a kitplane. Complete with photos and brutally truthful chronicles of the construction, each entry summarizes the challenges and triumphs.

Although the site doesn't get updated regularly (hey, they've got building to do), you can request auto e-mail notification when updates occur. When it's completed in a couple of years, the kitplane will evolve into a 220 mph sport plane. Read on about the Lancair and performance specs by clicking the hyperlink to Lancair.

Fee or Free: Free.

AVIATION ENTERTAINMENT

Virtual Horizons

http://virtualnorth.com/horizons

RATING

BRIEFING:
Expand your browsing horizons with an insider's account into the Canadian bush pilot.

Virtual Horizons—it's where the soaring spirits of Canadian bush pilots shower us surfers with cyber-collections of stories, articles, letters and photographs. Mostly navigating through the Great Canadian North, where even industry hardened pilots clamor for a window seat, some of Canada's best aviation writers and flyers give you a tour into their experiences.

There's talk and pics of beautiful landscape, memorable journeys with Cessnas, Beavers, and Otters. Feature stories span a trip into the high Arctic to flying a Cessna Caravan in West Africa. Although giant, high resolution pics run wild throughout the descriptive accounts, don't scroll past them unviewed. The download time might be a bit lengthy, but these romantic beauties deserve a peek.

The icing on the Virtual Horizons cake? Hints for best viewing and site navigation are plenty. Thoughtful tips on weathering the lengthy image loading waits will increase a beginning browser's enjoyment. And, descriptive links guide you with pinpoint accuracy before you click.

Fee or Free: Free.

AVIATION ENTERTAINMENT

From Buffalo to Alaska

http://www.vivanet.com/~jruh/09.htm

RATING

BRIEFING:

A Cessna 172, father and son at the controls, and twenty days of Buffalo-to-Alaska flying—sounds like the perfect recipe for online euphoria.

Warm. Humorous. Candid. Inspirational. Descriptions commonly echoed while congratulating aviation's finest authors do apply here as well. No, these guys aren't professional writers, nor cross-country regulars. We're simply invited to peek into the journal of a father-son expedition into the awesome spectacle that is Alaska. The tools: a Cessna 172, a laptop computer for journal entries, and a digital camera.

This online, amateur documentary is brought to life with capable daily descriptions and fantastic photos. Broken down and indexed into twenty days of unforgettable adventure aloft, this Buffalo-to-Alaska journal speaks of engine problems, Sunday afternoon repairs, treacherous weather, on-foot exploring, riding low pressure waves, "ugly VFR," Hurricane Fran, and more. Inline links throughout the entries give you a quick jump into extra details like maps and photos. Or, move through the entries undistracted with "next day/previous day" directions.

Whatever your favorite way to browse, this online journal captures the imagination and thrill of flying. I mean, how can you go wrong sauntering through places like Medicine Hat and Moose Jaw?

Fee or Free: Free.

AVIATION ENTERTAINMENT

Captain J's Aviation Page

http://www.geocities.com/CapeCanaveral/1274/contents.html

RATING
✈✈✈

BRIEFING:
Captain J has just turned off the "no-enjoying-yourself" sign. It's time for a romp in aviation distraction.

Getting down to serious diversion, Captain J's Aviation Page avoids the facts, the FARs, and the forecasts. Replacements take the form of more trivial, but worthwhile aviation endeavors like: movies, games, jokes, poetry, photos, and sounds. In fact, this fanciful virtual compilation offers nothing to make your flying safer, forecasting easier, or the regulations clearer. Captain J is here only for fun.

Certainly taking the low road to reach this shaky organization and design destination, Captain J shows concern solely for aviation entertainment—no matter the means. You'll forgive and forget, however, as you reach the series of online delights. Clicking into Sounds brings to life a 747 flyby, an F-16 on takeoff, San Francisco's ATIS (full 60 seconds), a jet taxiing, and more. Similarly presented, Airplane Movies, Aviation Games, and Photos are ripe for the clicking. Examples? Grab some popcorn for QuickTime movies like: The X-1 Released from a B-29. Avoid crashing your plane into the mountains with the Alaska Flight Game. And, scan through high resolution pics like a 747 at EAA Fly-In 1994. Then, maneuver your mouse into the inspirational poetry and anecdotes. They're equally inviting.

Fee or Free: Free, even if you have to download the plug-ins.

AVIATION ENTERTAINMENT

Flying Contraptions Home Page

http://www.prysm.net/~jnuts/home.html

RATING

BRIEFING:

A jumbled mix of unusual flight delights. Strap on a jet pack and soar with this collection of contraptions.

If you're reading this book, you'd probably agree that flight is a fascinating endeavor. The author of the Flying Contraptions Home Page thinks so too—he's chronicled a collection of devices invented to take flight. By including this site among 200 of aviation's best, you may believe my directional gyro's on the blink. Read on, however, for some entertaining insight into seat-of-your-pants gizmos.

For a taste of content, look no further than the site's keywords used for searches. Topics like rocket belt, individual lift device, one-man helicopter, Project "Grasshopper," and aerocars should be a strong clue as to the oddities included. The majority of concoctions researched here span the 1950–1970 time frame—with most never lifting past the prototype stage.

Each captivating article summarizes project histories, accomplishments, and statistics. Teaming up with photos and other research links, the articles succeed at serving up a bit of entertaining info while avoiding textbook style dreariness. Sure, the text is riddled with typos and punctuation blunders, but it doesn't matter. Hey, you're here to have fun, loosen up.

Fee or Free: Free.

AVIATION ENTERTAINMENT

Solo Stories

http://www.geocities.com/CapeCanaveral/3831

RATING
++

BRIEFING:
Sensational solo remembrances—sans instructor.

The fact that the Solo Stories site is updated continuously, easily navigated, and packed with stories has nothing to do with my overwhelming recommendation. The brutal honesty and unbridled realism is why I've cleared room for a bookmark.

Granted, Solo Stories brings back personal memories of my own solo and cross-country days, but I'm sure it will do the same for others. Whether you care to relive your solo adventures through the memoirs of others or learn from their mistakes before you fly, check in here for a taste of reality. For a cyber-ride of jittery, real-life flights, I suggest clicking into stories titled, "If You Run Into Traffic, Tell Them You're a Student on Your First Solo—They Will Get Out of Your Way;" "As I Latched the Door Behind Him, My First Words Were 'Oh Sh_ _';" and "When I Landed Safely, I Felt Like Amelia Earhart!"

Then, when you're ready to recount your experience, a link guides you into adding your own solo story. It's simple to contribute, just look at the many entries already showcasing their maiden voyage.

Fee or Free: Free.

AVIATION ENTERTAINMENT

Discovery Online—Wings Conversations

http://www.discovery.com/area/technology/technology.html

RATING
✈ ✈ ✈ ✈

BRIEFING:
Discover this popular flight forum's appeal of real-life recollection.

Creating the perfect lift for grounded flying fans, Wings Conversations lines up a healthy array of real-life stories and unedited thoughts. You'll find the world of aviation comes quickly into focus with these uninhibited tales from fellow flight friends. If you're more intrigued by fancies of flight over Aardvarks or Rock-em-Sock-em Robots, my suggestion is to skip the hordes of Discovery info, and take a shortcut with the long-winded address at the top of this page.

Your illustrious site editor provides the fuel for the posting frenzy with topics like: Debacles and Delights in Flight, Take Off From There!, World War II Flyboys Meet Here, and Your Favorite Stories. Passionate entries fill the pages, with some having a string of responses. Examples of the flowing madness include: Air Sick Pilot; 1 Inch of Clear Ice on a C-172, and I Walk Away; Inflatable Airplanes; The Legendary F-4 Phantom, and The Unknown Hero. You get the idea.

With hundreds of postings, the online stratofortress of conversation can easily fly off into an unmanageable fog. So, ground rules for posting do apply and are expressed clearly. Be sure you log them into memory before typing.

Fee or Free: Free.

200 Best Aviation Web Sites

AVIATION ENTERTAINMENT

Fudpucker Airlines

http://www.infinet.com/%7Ecuban8

RATING
✈ ✈ ✈

BRIEFING:
Frolicking in a hilarious haze, Fudpucker Airlines captures aviation's priceless perceptions.

Fee or Free: Free, unless you have an overwhelming urge to send them money as they request.

Way out in restricted airspace, cruising at oxygen altitudes, Fudpucker Airlines has obviously shunned the comfort of pressurization and opted for mind fog. Put another way, whenever the site navigation options are labeled "go back, go back!" or "live dangerously" you begin to realize your online airfare was cheap for a reason.

Just scanning the index of the Fudpucker Pages should give you a clue as to this site's comical content. All About Fudpucker; Ashley's Hot Tips for Cool Pilots; Where To Go; What to See; The Dudley P. Fudpucker Complete Guide to Aviation Terminology; Who Are Those Guys?!; and more pointless drivel you can't live without. Whatever your fancy, do remember to observe the online passenger rules. Among those of which you should be aware: "Don't get snooty with the crew;" and "Remember, your pilot is still learning to fly and he is more scared than you."

While you're online with Fudpucker, Night Flight must be among your site destinations. The countless stories (mostly true) are hilarious and ramble on for pages and pages. Simply put, these amazing tales are worth the bookmark space alone.

AVIATION ENTERTAINMENT

Paper Airplanes
by The PC Help Group

http://pchelp.inc.net/paper_ac.htm

RATING
+ + +

BRIEFING:
Perfect for the home, office, or outdoor adventure, these gliders make your childhood spirits soar!

Remember when grandpa expertly added those taped ailerons to your simple paper airplane? Well, I'm sure a lot of aviation wisdom and engineering dreamers came up with this little jewel of a site.

Although under some updating at the time of review, the page and what it offers just screams, "STOP WHAT YOU'RE DOING AND HAVE A LITTLE FUN!" From an Origami Aerobatic Design to a Supersonic Fast Flyer to a Soaring Glider, anyone running the age spectrum will delight at these propeller-less paper creations. There's a design for every month to try. Step by step diagram instruction glides you through assembly.

It really is simple, free fun. Click, print, fold, and enjoy!

Fee or Free: Free.

AVIATION ENTERTAINMENT

"Dad" Rarey's Sketchbook
Journals of the 379th Fighter

http://www.rarey.com/sites/rareybird/index.html

RATING

BRIEFING:

An illustratively chronicled tribute to a WWII fighter pilot that deserves a look.

Painful at times and uproarious at others, this wonderful hidden gem of a site delves into a personal account of Mr. George "Dad" Rarey. Drafted into the Army Air Corps in 1942, this young cartoonist and commercial artist kept an animated cartoon journal of the daily life of the fighter pilots. Brought to the bookshelves and now cyberspace by his son and wife, this thoughtful reflection chronicles "Dad" Rarey's WWII life.

Skillfully prepared and graphically rewarding, this home page tribute has all the stuff that come under the heading of great organization: excellent page links, clickable menu icons throughout, and tiny thumbnail pics that don't waste time (and can grow at your command). But, by far, the best page features are the written descriptions and illustrations. Contributions you find here come from surviving members of the 379th Fighter Squadron, excerpts from Rarey's letters to his wife (Betty Lou), and Betty Lou's memoirs.

Clickable sections include: Cadet Life, Volumes 1–5, Nose Art, and Artifacts.

Fee or Free: Free.

AVIATION ENTERTAINMENT

Aviation Jokes

http://www2.mcis.duke.edu/CAP/Misc/Jokes.html

RATING
✈ ✈

BRIEFING:
Non-graphically driven collection of flying flippancy and aeronautical amusement.

Quite frankly, I'm not sure how often or to what extent this humor driven home page is updated with fresh stuff. However, the page "Augmentor" assures us that newest jokes will secure a coveted place at the bottom of the page so "repeat readers can find them in the pile." Whatever the update status, this page does deliver on fabulously funny flying fancies and the like. Albeit the page esthetics take a winning position among the yawning variety, you'll forgive and forget as you read on.

Without revealing too much about these treasures, expect to stumble across stuff relating to: God and pilots, Santa Claus and an unexpected FAA inspection, student/instructor high jinks, the greatest lies in aviation, an accident eyewitness account, and other aviation anecdotes.

The Aviation Jokes page uncovers a hodgepodge of lists, one-liners, article excerpts, dialog stories, and questions/punch lines—all from a variety of sources. Even without any idea of page longevity, this site's recommended for frivolous flying fun. Take it in while it lasts!

Fee or Free: Free.

AVIATION ENTERTAINMENT

Bookmarkable Listings

Air Pix Aviation Photos
http://www.cincymall.com/airpix
Collection of aviation photos and related for-sale products.

Tom Claytor—Bush Pilot
http://www.mck.co.za/bushpilot
Follow the travels of a bush pilot on a solo expedition to the seven continents of the world.

Greg's Common Commercial Aircraft Spotter's Guide
http://www.geocities.com/CapeCanaveral/1273/spotting.html
Brief descriptions and photos identify two-engine, three-engine, and four-engine commercial aircraft.

The Spotter's Homepage
http://www.IAEhv.nl/users/dirkx
Netherlands-based site offering information and links for worldwide aircraft spotting.

Aviation Employment

Aviation Employee Placement Service (AEPS)

http://www.aeps.com/aeps/aepshm.html

RATING
☦ ☦ ☦

BRIEFING:
Get yourself in front of over 600 aviation companies with a few clicks of the mouse.

Solely text-based and unconcerned with visual niceties, The AEPS still soars with their worthwhile, well-organized employee site. Billed as the "online job connection," AEPS offers an award-winning selection of resources built upon its main menu of: New Visitors and Inactive Members Page, Aviation Companies (for employers), and Other Info.

Once an active member, you are invited to partake in a long list of menu options including: Renew Your Membership, Member Aviation Companies (over 600 at time of review), Update or Add Your E-mail (keep e-mail current to receive The Aviation World Reports, job alerts, etc. automatically), Update Your Qualifications (all updates are free), View Your Qualifications, Conduct a Sample Search, Aviation Info Exchange, AEPS Newsletter, AEPS Feedback, and of course, The Active Members Jobs Page.

As always, job postings and data bank searching are free for all employers.

Fee or Free: Free.

AVIATION EMPLOYMENT

Air, Inc.—The Airline Pilot Career Specialists

http://www.airapps.com

RATING

BRIEFING:
Plan a thorough career course here before you go wheels up.

For airline career seekers Air Inc.'s lofty online resources span the industry to give you a jumbo-sized heads-up. Climb aboard an unrivaled career guide for pilots.

Without so much as a mouse click, the latest airline industry tidbits are ready for the reading in Have You Heard. Once you've caught the latest hirings and dealings, a left-margin list of topics are close at hand: Resource Center, A.P.C. Magazine (online feature articles from the printed sibling), hiring summary, airline profiles (extremely in-depth), career counseling, resume service, seminar and job fairs, calendar, publications, and more.

Where you'll probably spend the most time is in the Resource Center. Read countless articles broken down by category: Airline Pilot Center, Military Pilot Center, Low-Time Pilot Center, and Student Pilot Center. Nifty site additions like automatic e-mail updates, contest prizes, and a fantastic bulletin board help propel Air, Inc.'s Web wonder nautical miles past the rest.

Fee or Free:
Mostly free. Worthwhile career services and magazine subscription are fee-oriented.

200 Best Aviation Web Sites   **229**

AVIATION EMPLOYMENT

Aviation Jobs Online

http://www.aviationjobsonline.com

RATING

BRIEFING:
Award-winning seeker site. Yes it's fee-oriented, but they do all the grueling work.

Searching for that dream job sometimes requires as many allies as one can muster. Sign up with Aviation Jobs Online and you'll instantly have the beginnings of a powerful job search—24 hours a day, seven days a week.

Self proclaimed as "the only site on the Internet that maintains a current airline directory which includes minimum qualifications," Aviation Jobs Online is your complete employment source. Although site navigation leaves a bit to be desired (plan on lots of scrolling), informative topics abound. There are free and membership-only areas, special offers and contests, a Book Store, AvJobs Business Directory, Aviation News Now, free job posting for employees, and more.

After you get through the hard sell areas, you'll find that these folks are serious about aviation jobs with their resourceful personnel service. They'll help with resumes, finding articles on specific companies, and help to prepare you for the job interview.

Automatic e-mail updates, over 2000 links, Zip Code Weather, and Electronic Post Office round out the site's cool creations.

Fee or Free: Fee-related. Many fee options are available—see site for details.

www.FindAPilot.com

http://www.findapilot.com

RATING
++

BRIEFING:
Meet the no-frills, yet focused employment matchmakers at www.FindAPilot.com.

Fee or Free: Fee-oriented services available.

Relatively new at the time of this review, www.FindAPilot.com takes the active with its own cyber-version of online employment exchanges. The site seems to avoid any gratuitous visual pleasantries—skipping right to the meat of the matter. It's all about jobs and aviation pros. Period. No dreary news. No QuickTime flybys or audio oddities. Serious aero hunters will appreciate the fluff-free focus. It's refreshing.

A similar text-based menu springs up everywhere, giving even frantic job searchers easy maneuverability. The occupation-only offerings for position shopping pilots include a jobs listing area, where employers are invited to post free listings; FAQs and more information; and fee-oriented services of resume posting. Personal "home page resumes" include your choice of background colors or wallpaper, multiple category listings, user defined links, and free unlimited updates!

Employers shopping for new recruits will enjoy free position posting and handy pilot searching tools. View the resume database (although a tad thin as of review time, watch for a quick increase in volume) alphabetically or sort by job type. Listings are categorized into: airline, avionics tech, cockpit crew, corporate, flight instructor, ground crew, helicopter, management, mechanic, reservation/ticket agent, seaplane and others.

AVIATION EMPLOYMENT

Airline Employment Assistance Corps (AEAC)

http://www.sni.net/AEAC

RATING
+ + +

BRIEFING:
A resourceful fee-oriented aviation career counselor. You'll find (or fill) that long-awaited aviation position with help from the AEAC people.

You've hung out long enough down at your local airport. Face it, you'll need to get a little more serious if you're going to find an aviation job. But, hold on to your headset, the Airline Employment Assistance Corps is your new online resource. Although the name would imply airline only, there's room here for any career in aviation.

This lofty employment service provides a long list of helpful categories: worldwide classified ads; resume resources (post your own here, or get resume help from pros); an industry look at opportunities (titles, salary ranges, education requirements, and employers); airport careers; aviation and maintenance careers; air traffic controllers; aviation safety inspectors; flight attendant careers; government aviation careers; pilots and flight engineers; salary relocation calculator; and a host of aviation-related links.

Because this is a professionally maintained employment site, a membership fee is required. If you're really an aviation job seeker, the value here is obvious.

Fee or Free: Some free areas, but membership gives you access to everything.

Your Career in Aviation
The Sky's the Limit
http://www.tc.faa.gov/ZDV/careers.html

RATING
+++

BRIEFING:
Break into an aviation career armed with necessary info found here.

An extension of the FAA Aviation Education Program, this informative site gives you answers to your deepest, darkest aviation career questions. Start by picking a career: pilots and flight engineers, flight attendants, airline non-flying careers, aircraft manufacturing, aviation maintenance and avionics, state aviation jobs, and more. Next, get a briefing on burning topics like: what's involved, pay, hours, working conditions, future outlook, and yes, where to track down your dream job.

Warning: information here is refreshingly candid—it's not all rosy and fluff. You'll gain honest insight into real-world requirements, conditions, and industry outlook. Start here before taking the active in aviation.

Fee or Free: Free.

Aviation/Aerospace Jobs Page
(NationJob Network)

http://www.nationjob.com/aviation

RATING

BRIEFING:
Looking for aviation employment? This FREE, professional service does the job.

Offered up by the huge employment resource, NationJob Network, aviation opportunities abound here. Access an endless sea of jobs one of two ways: either click on home page logo icons for some big name industry leaders (Boeing, Learjet, Raytheon Aircraft, Cessna, The Nordam Group, Thomson Saginaw, etc.), or search through jobs listed here by location, position type, salary, keyword, and more. Simply click on an appealing job in the lineup. From there, you gain access to a company profile as well as an enlightening job description. Most likely you'll be overwhelmed by the variety of categories. Positions range from flight test engineers to buyers, and from A&P service mechanics to vice presidents of operation.

Even if you're not thrilled with the arduous task of sifting through these nationwide listings, just ask "P.J. Scout" to do it for you automatically. This convenient little feature makes employment hunting effortless with an e-mail notification service. Simply enter your job preferences and e-mail address. "P.J. Scout" will search the furthest reaches of the Web and find jobs that match your parameters. He reports to your e-mail weekly. It's free, confidential, and cool.

Fee or Free: Free.

AVIATION EMPLOYMENT

Bookmarkable Listings

AeroTrek
http://www.fivesticks.com/aerotrek
Contact information database of over 5,500 operators from major airlines to small freight companies.

The Corporate Aviation Resume Exchange
http://scendtek.com/care
Employment bulletin board of resumes for corporate aviation professionals.

Universal Pilot Application Service
http://www.upas.com
Extensive pilot database offers exposure to position-seeking pilots and qualified pilot info to employers.

Airline Pilot Job Update
http://www.flyingjobs.com
Promo site for the monthly airline hiring newsletter, *Airline Pilot Job Update*.

Index

A

Aerobatics 29, 55, 63, 164
Aeronautics 34, 44, 53, 60, 63, 67, 138, 144, 173, 175, 191
Air charter 13, 36, 90, 103, 113, 153, 194
Air traffic control 11, 40, 47, 72, 122, 173
Airlines 8, 9, 14, 24, 37, 66, 69, 71, 97, 110, 113, 117, 123, 143, 155, 159, 163, 164, 226, 229, 235
Airshows 12, 19, 22, 33, 37, 41, 50, 52, 55, 56, 59, 62, 157, 162, 164, 174, 177, 178
Airspace 94, 117, 121
Airworthiness Directives 99
Antique aircraft 38, 144
Army 38
Aviation
  accidents 43, 58, 66, 98, 159, 177
  general 24, 31, 44, 63, 66, 93, 152, 168, 170
  images 18, 19, 22, 23, 24, 34, 106, 118, 136, 141, 155, 157, 162, 189, 190, 213, 226
  jokes 225
  law 94, 156, 158, 164
  maintenance 13, 48, 69, 71, 72, 99, 171, 194, 198, 206
  medicine 36, 51, 70, 98, 100, 156, 172
  policy 47, 57, 59, 65
  quotes 211, 214
  women in 35, 41, 42, 139, 144

B

Beechcraft 72, 134, 203

C

Careers 11, 42, 45, 46, 53, 118, 135, 138, 229, 232, 233
Cessna 25, 48, 109, 134, 137, 152, 210, 216, 217, 234

E

Earhart, Amelia 35, 54, 132, 139, 140
Employment 53, 71, 99, 122, 156, 202, 204, 228, 230, 231, 232, 234, 235
Experimental aircraft 44, 52, 60, 62, 63, 67, 114

F

Flight planning 24, 112, 113, 135, 164, 172, 180, 194, 198
Flight simulators 28, 38, 61, 111, 113, 122, 162, 177, 192, 195
Flight training 13, 23, 111, 117, 121, 122, 133, 135, 137, 142, 143, 158
Floatplane 153
Fly-in 59, 62, 68, 108, 154, 164, 167
Fuel 10, 23, 27, 91, 93, 113, 121

G

Glider 142
Global Positioning System (GPS) 24, 101, 104, 114, 177, 185

H

Helicopter 24, 36, 50, 72, 127, 167, 183, 204
Homebuilt 114, 149, 153, 154, 177, 183, 215

M

Meteorology 15, 80, 84, 88
Mooney 72, 152

N

Navigation 27, 34, 40, 73, 98, 121, 180, 206
Navy 55, 72

P

Piper 205

R

Radio-controlled 30, 36, 49
Restoration 113, 174

S

Safety 10, 43, 58, 65, 97, 98, 100, 107, 119, 133, 148, 155, 159, 165, 170, 172, 173, 178
Soaring 64, 142
Space 17, 44, 53, 129, 130, 132, 144, 160, 173, 175, 176, 178, 191, 204

U

U.S. Air Force 46, 56, 72, 73, 126, 129, 145, 166, 191

W

Women in aviation 35, 41, 42, 139, 144

About the Author

John A. Merry is a VFR-rated private pilot and has had an ongoing editorial column in *Plane & Pilot News*. He is a member of the AOPA and Pilots International Association and the owner of Specialized Marketing Agency, serving aviation-related companies with marketing consultation.